Images of scientists and ideas about science are often communicated to the public through historic biographies of eminent scientists, yet there has been little study of the development of scientific biography. *Telling Lives in Science* brings together a collection of original essays which explore for the first time the nature and development of scientific biography and its importance in forming our ideas about what scientists do, how science works, and why scientific biography remains popular.

Written in an accessible style by leading historians of science, several of them biographers, the volume explores such questions as how scientific heroes are fashioned for popular consumption, what relations exist between scientific biography and autobiography, and the role of biography in the history of science and medicine. The book comprises theoretical and historical studies and ranges from the seventeenth to the twentieth century, paying attention to such icons as Michael Faraday, Charles Darwin, Humphry Davy, Florence Nightingale and Sir Joseph Banks, along with a gallery of lesser-known figures. In their introduction the editors examine the development of scientific biography, its relations with literary biography, and consider some of the issues at the heart of contemporary debates in the literary and historical study of biography.

With its broad sweep and careful, imaginative scholarship, this volume provides a timely and challenging examination of an important aspect of the culture of science that will be of special interest to historians of science, academics and students, and the general reader interested in the popularisation of science.

Telling lives in science: essays on scientific biography

Telling lives in science

Essays on scientific biography

EDITED BY

Michael Shortland
Unit for History and Philosophy of Science, University of Sydney

and

Richard Yeo
School of Cultural and Historical Studies, Griffith University, Brisbane

CAMBRIDGE
UNIVERSITY PRESS

Published by the Press Syndicate of the University of Cambridge
The Pitt Building, Trumpington Street, Cambridge CB2 1RP
40 West 20th Street, New York, NY 10011–4211, USA
10 Stamford Road, Oakleigh, Melbourne 3166, Australia

First published 1996

Printed in Great Britain at the University Press, Cambridge

A catalogue record for this book is available from the British Library

Library of Congress cataloguing in publication data

Telling lives in science: essays on scientific biography/edited by
   p. cm.
Includes index.
ISBN 0 521 43323 1 (hc)
1. Scientists – Biography.   2. Science – History.   3. Scientists –
Biography – Authorship – History.   I. Shortland, Michael.   II. Yeo,
Richard R., 1948–   .
Q141.T45 1996
809'.93592'0245 – dc20   95-47399 CIP

ISBN 0 521 43323 1 hardback

Contents

Contributors

GEOFFREY CANTOR is professor of the history of science at the University of Leeds and is currently working in the area of history of science and religion. His biographical study of Faraday, *Michael Faraday: Sandemanian and Scientist*, was published in 1991.

Address: Department of Philosophy, University of Leeds, Leeds LS2 9JT, UK.

JOHN GASCOIGNE is a senior lecturer in the School of History, University of New South Wales, Sydney. His most recent work is *Joseph Banks and the English Enlightenment: Useful Knowledge and Polite Culture* (1994) and he is presently completing a companion volume entitled *Science in the Service of Empire: Joseph Banks, the British State and the Uses of Science in the Age of Revolution*.

Address: School of History, University of New South Wales, PO Box 1, NSW 2033, Australia.

MICHAEL HUNTER is professor of history at Birkbeck College, University of London. He has written or edited various books on the early Royal Society and on intellectual change in the late seventeenth century. His *Robert Boyle by Himself and his Friends* was published in 1994. With Edward B. Davies and Antonio Clericuzio, he is currently preparing a complete edition of the works and correspondence of Robert Boyle.

Address: History Department, Birkbeck College, University of London, Malet Street, London WC1E 7HX, UK.

DAVID KNIGHT is professor of history and philosophy of science at the University of Durham. He is currently President of the British Society for the History of Science. He is general editor of the series of Cambridge's Scientific Biographies, in which his own *Humphry Davy: Science and Power* is published. He has written widely, especially on the history of chemistry.

Address: Department of Philosophy, University of Durham, 50 Old Elvet, Durham, DH1 3HN, UK.

JAMES MOORE teaches history of science and technology at the Open
University in Milton Keynes. His and Adrian Desmond's *Darwin*
(1991) won the 1991 James Tait Black Memorial Prize for
Biography and the 1993 Watson Davies Prize of the History of
Science Society. His study *The Darwin Legend* was published in
1994. He continues to research on Victorian science and religion,
Alfred Russel Wallace and twentieth-century popular science.
Address: The Open University, Walton Hall, Milton Keynes MK7 6AA,
UK.

DORINDA OUTRAM teaches history at University College Cork,
and is a Visiting Scholar at the Max-Planck-Institut für
Wissenschaftsgeschichte, Berlin, 1995–1996. She is the author of
Georges Cuvier (1984); co-editor of *Uneasy Careers and Intimate
Lives: Women and Science 1789–1987* (1987), and *The Body and
theFrench Revolution* (1989).
Address: Max-Planck-Institut für Wissenschaftsgeschichte, Wihelmstrasse
44, Berlin D-10117, Germany.

ROY PORTER is professor of the history of medicine at University College
London, and has written extensively on the social and cultural
history of science and medicine. Amongst his many books are
studies of Edmund Gibbon and Thomas Beddoes, *English Society in
the Eighteenth Century* (1982), *Mind Forg'd Manacles* (1989) and
London: A Social History (1994).
Address: Wellcome Institute for the History of Medicine, 183 Euston
Road, London NW1 2BN, UK.

MICHAEL SHORTLAND is associate professor in the history and
philosophy of science at the University of Sydney. His recent
writings have been concerned with the social and cultural history of
nineteenth-century science, and include his edited collection *Hugh
Miller and the Controversies of Victorian Science* (1996).
Address: Unit for the History and Philosophy of Science, University of
Sydney, Sydney, NSW 2004, Australia.

THOMAS SÖDERQVIST is associate professor at Roskilde University,
Denmark, where he teaches history of science. He has written on

the history of ecology (*The Ecologists* 1986) and on various problems in the historiography of science. He is now completing a biography of the leading theoretician in post-war immunology, Niels K. Jerne.

Address: Department of Biology, Roskilde University, PO Box 260, DK 4000, Roskilde, Denmark.

MARTHA VICINUS is Eliza M. Mosher Distinguished University Professor of English, History and Women's Studies at the University of Michigan, Ann Arbor. She has written or edited numerous articles and books on Victorian women and on the history of sexuality, including *Ever Yours, Florence Nightingale: A Selection of Letters* (1989), co-edited with Bea Nergaard. She is currently writing a book on the formation of modern lesbian identities, 1780–1930, *Perverse Women*.

Address: English Department, University of Michigan, Ann Arbor, MI 48109, USA.

RICHARD YEO is associate professor in the history and philosophy of science at Griffith University, Brisbane. He has written widely on eighteenth- and nineteenth-century social and intellectual history of science and co-edited *The Politics and Rhetoric of Scientific Method: Historical Studies* (1986). His *Defining Science: William Whewell, Natural Knowledge and Public Debate in Early Victorian Britain* was published in 1993.

Address: Faculty of Humanities, Griffith University, Nathan, Brisbane, Queensland 4111, Australia.

Preface

Biography has for long been an important medium in the transmission of images of scientists and ideas about science. Whether in the form of celebratory accounts of the heroes and heroines of science or more rounded, critical studies, the lives of such famous figures as Newton, Darwin, Einstein and Madame Curie have always been assured a ready market and an eager audience. Today, as seldom before, the general public is hungry for news about science and consumes scientific biography with relish.

Perhaps surprisingly, very little criticism or comment has accompanied the recent resurgence of interest in scientific biography; nor are there readily available any detailed or comprehensive studies of the development of the genre, its status and influence. *Telling Lives in Science* is in fact the first book of its kind.

Since the seventeenth century, memoirs, anecdotes and portraits of leading natural philosophers have been central in forming the way in which the scientific community is perceived by outsiders and perceives itself. Although the history of science has blossomed over the past twenty years and undertaken detailed explorations of the culture of science, biography has until recently been viewed as an old-fashioned, stale and distinctly uninteresting resource. Far more attention has been paid to scientific textbooks, novelistic tales of science and laboratory notebooks than to biographical accounts. While the former, properly interpreted, can be immensely rich and revealing, the latter have been often dismissed as simplistic and self-serving. Yet, the promise of biography, to tell the truth about a life, is surely worth taking seriously. So, too, is its occasionally brilliant accomplishment in the integration of a historical subject's private, even intimate, life of feeling with the broader, public domain of career and honour.

That this is coming to be more widely recognised is indicated by a number of seminars and conferences devoted to scientific biography and held over the past three years in Canberra, New Orleans, Toronto and London. The nature and role of scientific biography are questions today possessing a certain urgency, the result perhaps of the project underway to prepare a *New Dictionary of National Biography*. It has long been recognised that the existing DNB, a magisterial publishing enterprise that capped the nineteenth century's enthusiasm for the lives of great men, is not only out of date but deficient in its coverage of many scientific figures. The prospect of writing a new 'Michael Faraday' to replace John Tyndall's is challenging enough; more daunting is the necessary task of integrating in a succinct account the mass of scholarship that has appeared since Richard Glazebrook wrote his compendious, but disorganised, entry on Isaac Newton. We do not suppose that the essays in this collection will serve as a crib for the new band of scientific biographers charged with composing 'lives' that will be read in the next century. Nevertheless, we expect that they and other readers will find in this book many new lines of enquiry and promising perspectives for any reassessment of scientific biography.

In preparing this volume, Michael Shortland and Richard Yeo have received the support, respectively, of the Unit for the History and Philosophy of Science at the University of Sydney and the Faculty of Humanities at Griffith University. For their assistance in providing help and information, we also thank Shari Lee, Malcolm Oster, Jon Wennerbom, Karen Yarrow and Jo Wodak. The contributors, and Fiona Thomson, our editor in Cambridge, have provided advice and shown patience throughout the preparation and production of this book. *Telling Lives in Science* recalls Marc Pachter's *Telling Lives: The Biographer's Art* (1979), a study whose range and interest, as well as punning title, will we hope be complemented by our own.

Michael Shortland
Richard Yeo

Introduction

MICHAEL SHORTLAND AND RICHARD YEO

Biography today occupies an unusual, perhaps even an uncomfortable, place in our culture, being one of the most popular and yet least studied forms of contemporary writing. Readers in their tens of thousands consume biographies avidly and offer publishers a sure market for the life stories of writers, musicians, film stars and sporting heroes. 'The Age of Biography is Upon Us' announces a recent article (Bowker 1993), as anyone will know who steps into a bookshop or glances at the year's best-seller lists. A 1994 poll on reading habits in Britain showed biography to be the most popular category of non-fiction book, selected as their favourite by 19 per cent of readers, a number matched by the most popular category of fiction, 'romance', and considerably ahead of contemporary fiction, read by just 14 per cent of readers (D.S. 1994; see also Beauchamp 1990 for US data). The market for biography is by no means restricted to the 'popular' end of the spectrum: a typical issue of the *Times Literary Supplement* (21 October 1994) carried a lead review devoted to a biography of William Tyndale, an extended review of a biography of Mrs Dorothy Jordan, and reviews of biographies of Kremlin wives, and Nikolai Bukharin's widow: in its listings, there were more entries under 'Biography' than under 'Fiction', 'Politics', 'Literature and Criticism', more than in any other category. 'Biography' has become a potent selling tag for all manner of books that in a different age would have been shelved alongside works of history, geography, natural history and bibliography: subtitles such as 'biography of a book', 'biography of a small town, 'biography of a breed', 'biography of Miami Beach', 'biography of a tree', 'biography of Victorian marriage', 'biography of an eye' and 'biography of England' now aspire to the sales achieved by biographies of human beings, alive, dead, famous or unknown (Epstein 1987: 82–3; also Skidelsky 1987).

Biography has become so common that one is compelled to consider whether there is not, now, too much of it. With nearly 250 biographical studies of Samuel Johnson, 60 or so of Charles Dickens and over 75 of James Joyce, the question has merit. Smaller academic industries are devoted to producing commentaries and biographies on major scientists such as Isaac Newton, Charles Darwin and Albert Einstein; consequently, Darwin is memorialised and celebrated with editions of his volumes of notebooks, with a massive project to publish all his known correspondence and, in support, his diaries and editions of his collected and single writings (in first, variorum, facsimile, abridged, illustrated and collector's editions) and, since 1990, three substantial biographies with a fourth, projected to run to two volumes, on its way. A volume has even appeared (with another promised) comprising Darwin's annotations to his own books. Under these circumstances, one can sympathise with the biologist Richard Lewontin who set out to review yet another biographical addition by noting with some exasperation that his university library already held over 180 books about Darwin's life and work (Lewontin 1985, p. 18). The facts have surely now been established beyond doubt — perhaps to the point of monotony. Are any further biographies really needed, except to satisfy a public's apparently unquenchable thirst?

There is, in fact, always room for more biographies of Darwin, but we say this with some measure of reluctance and even some misgiving (and, since we publish in this book an illuminating 'metabiographical' reflection by James Moore of his experience in co-writing the acclaimed *Darwin* (1991), some hesitation as well). While new studies can and should serve the useful function of correcting errors and filling gaps in earlier accounts, not all have done so, and the worthwhile discovery of archives, letters and materials is not always integrated into new biography. Leaving this aside, new studies are needed as each new age redefines itself, and fresh questions come consequently to be asked of biography. Our moral and epistemological beliefs shift and our assessments of personality and behaviour follow suit. So, too, do our ideas about what biography is, can be, and should do.

So much for the prescription; but scientific biography has not always registered these sea changes. Many biographies still lie, so to speak, on the beach, untouched by philosophical and historical developments, offering stale narratives of heroism, selflessness and devotion to duty. As for our misgiving, this arises from the fact that as the biographical spotlight is turned again on Charles Darwin or Charles Dickens, other, equally fascinating figures remain in the shadows: myriad biographical Darwins, but only one modern biography of his 'bulldog' companion, T.H. Huxley. To date, dozens of other prominent scientific figures from the period have received no serious and sustained biography whatsoever: Alfred Russel Wallace, William Buckland, Robert Chambers, John Tyndall, Adam Sedgwick, David Brewster, to name but a few.

Those who write about biography are in the main those who write biography, and it is still comparatively rare to find professional critics or historians at conferences on biography or in the columns of newspapers and literary reviews. Over half the participants at a landmark meeting in 1988 were professional biographers not academics (Charmley and Homberger 1988). Professional biographers ask questions about biography that fit uneasily with the concerns of the modern academic community. Evoking a range of traditional or personal motives, biographers have been concerned with the usefulness and truthfulness of biography, with its status as art and literature, and with its claims to be able to present an authoritative or even definitive life. They speak and write unselfconsciously about 'interesting' lives, 'telling good stories', 'righting a wronged reputation', and about how best to evoke personality. Critics, meanwhile, hardly consider biography at all and, when they do, it is in terms of narrative, rhetoric and discursive structures. A glance at many recent books and articles of literary criticism suggests that the very terms with which biographers set about their work have been put into question (and scare quotes): 'truth', 'authenticity' and 'definitive'. Not surprisingly, even the most conscientious of modern biographers consider themselves liberated rather than impoverished by the lack of attention their work commonly receives from critics. 'We are blessedly free and untrammelled', writes Richard Holmes, author of biographies of Shelley, Coleridge and Savage, in contrast to novelists who 'are now doomed to work within a roaring factory of academic criticism, a ceaseless chatter of thesis and theory, and the distant pounding of the Great Tradition' (Holmes 1990: 21). Leon Edel, a rare combination of biographer and critic, declared in 1981 that no biographical criticism worthy of the name exists, that is to say, no criticism that connects with the interests and concerns of the writers (and readers) of biography (Edel 1981: 10).

What makes this gulf perplexing is that biography has recently emerged, or rather reemerged, as a genre of writing with considerable similarity to the novel, a subject of perennial interest to modern criticism. It is not only that biographies have begun to capture a general readership, once exclusively associated with 'Fiction', and that biographies are often displayed alongside novels, but that some of the once hard-and-fast distinctions between the two have broken down (see Kaplan 1994). Many distinguished writers have recently been drawn across the boundaries between fiction and biography — one thinks of A.N. Wilson and Peter Ackroyd, and of the superlative novels by John Banville on Copernicus (1976) and Kepler (1981) — while novelists have become fascinated with the activities of the biographer at work, as witnessed in Julian Barnes' *Flaubert's Parrot* (1984) and in Antonia Byatt's *Possession* (1990), a beguiling tale in which hero, heroine and villain are all conscience-tortured biographers on the trail of truth.

More germane to readers of this collection is the large gap between working biographers and historians. For example, our widening knowledge of the Victorian era has not yet made a profound mark on biographies of Victorian men and women. Biographers excel at the evocation of Victorian personalities — even Queen Victoria herself, in Elizabeth Longford's sensitive study — without taking much account of the broader historical trends and meanings. Indeed, a recent 'intimate biography' of Victoria by Stanley Weintraub (1987) is more preoccupied with the Queen's medical details, with sex, pregnancy and depression than with any public activities like wars, elections and Acts of Parliament. Except for the distinctively nineteenth-century ring to her ailments, she might have suffered in the late twentieth century instead of the late nineteenth century.

Biography and the history of science

In the early twentieth century history of science was still largely informed by a biographical approach: for example, Philipp Lenard's *Great Men of Science: a History of Scientific Progress* (1933) (see also Ireland 1962; Kragh 1987: 168). But as history and philosophy of science became professional academic pursuits — roughly after 1940 — most of the influential theoretical frameworks relegated biography to the sidelines. The positivist philosophers of science were not interested in the process of *discovery*; they regarded such speculations as fraught with psychological pitfalls and argued, instead, that the defining features of science lay in the rigorous method by which it verified results and tested theories. One consequence of this approach was a view of the history of science as an steady accumulation of knowledge; particular truths added to a larger edifice of established truths. In 1865, the French physiologist, Claude Bernard, who accepted the more critical parts of Auguste Comte's positive philosophy, anticipated this outlook: 'in this fusion [of successive truths], the names of the promoters of science disappear little by little, and the further the advances of science, the more it takes an impersonal form and detaches itself from the past' (Bernard 1957: 42, cited in Daston 1992: 613). The assumptions here were interestingly betrayed in the remark of H. W. Carr, the biographer of Gottfried Leibniz: 'The history of philosophy is essentially biographical. We cannot dissociate the philosopher from his system in the same way that we are able to dissociate the scientific discoverer's discovery from the scientific discoverer himself' (Carr 1929: 203, cited in Hankins 1979: 9). Of course, the historiography of science that followed in the years after Carr's observation – such as the work of Edwin Burtt, Alexandre Koyré, and E. J. Dijksterhuis – did not abstract scientific thought so brutally from its intellectual context, but in its concern with deep metaphysical structures, it did not encourage biography.

The sociology of science had similar implications for biography. Robert K. Merton's work began within the ancient genre of prosopography – or collective biography in the tradition of Plutarch, and later Francis Galton and Alphonse de Candolle (Galton 1869; Candolle 1873; Mikulinsky 1974). Merton correlated the beliefs and attitudes of a group of seventeenth-century English Puritans who pursued scientific interests (Merton 1938; on prosopography, see Shapin and Thackray 1974; Pyenson 1977; Kragh 1987: chapter 16). But his subsequent work, and that of the sociological programme he initiated, demoted any close study of individual scientists in favour of an analysis of the social norms and conventions governing the scientific community (Merton 1973). Thus since the 1940s a concern with the 'merely personal' has been seen as peripheral to the aim of understanding the scientific enterprise as a collection of institutionally based cognitive disciplines. From the 1970s, scholarly developments in the social history of science and sociology of scientific knowledge, and inquiries indebted to either Thomas Kuhn or Michel Foucault, have confirmed this situation.

The result has been that, in contrast to the case of literary biography, critical commentary on scientific biography has had to begin with a *defence* of the enterprise. This is the task Thomas Hankins accepted in his essay of 1979 which sought to bring biography in from the cold. Hankins asked historians of science to reconsider biography as the close testing point for broader theories of scientific thought and progress. Another affirmation came from Larry Holmes who saw biography as the entry point for any study of the 'fine structure of scientific creativity' (Holmes 1981, also 1974). But as Susan Sheets-Pyenson recently noted, Hankins was wary of the difficulties of integrating the non-scientific parts of a subject's life into the kind of intellectual biography he recommended (Sheets-Pyenson 1990). This continues to be a difficult issue: Pearce Williams, who wrote a biography of Michael Faraday in 1965, recently expressed doubts about a comfortable marriage between biography and any strongly sociological history of science, suggesting that the latter approach puts blinkers on our study of great and complex individuals, limiting what a detailed biography might discover. Citing a worst case scenario, he referred to Bruno Latour's *The Pasteurization of France* (1988) as 'Hamlet without the Prince of Denmark' (Williams 1991: 210; for other comments on biography, see Zoubov 1962).

Over the last decade or so there does seem to have been a resurgence of interest in scientific biography among historians of science (for a recent comment see Garfield 1990, 3). Notable examples of the genre include Dorinda Outram's *Georges Cuvier* (1984), Richard Westfall's *Never at Rest: a Biography of Isaac Newton* (1980), Thomas Hankins's *Sir William Rowan Hamilton* (1980), Dorothy Stein's *Ada* (1985), Walter Moore's *Schrödinger* (1989), Crosbie Smith and Norton Wise's *Energy and Empire: a Biographical Study of Lord Kelvin* (1989), Geoffrey Cantor's *Michael Faraday* (1991), Adrian Desmond and James Moore's *Darwin* (1991), Arthur Donovan's *Lavoisier* (1993), James Gleick's *Genius: the Life and Science of Richard Feynman* (1992) and Stephen Gaukroger's *Descartes* (1995). Some cautiousness about biography as the professed form of inquiry is still noticeable in some of the full titles of these and other works, which make it clear that the person is a key to some larger theme (see also Geison 1978; Yeo 1993).

Nevertheless, to different degrees, these authors seem to agree that biography, informed by investigation of primary sources and painstaking chronological detail, can yield the integration of intellectual and institutional narrative, of cultural and economic life, that is now valued in social and historical studies of science. If this is so, it suggests a shift from the earlier view in which biography was usually regarded as a point of resistance to these more general accounts of the Nature of science. Some recent interest in collective biography might be another sign of change (see Elliott 1982 and 1990; Abir-Am 1991).

This book takes its cue from these developments and offers a set of chapters by historians who consider a large span of questions about scientific biography. They do this from a variety of perspectives: some from a reflection on their own experience as biographers; others from an awareness of the role of biographies as primary material for historical study; others still from an interest in the status of biography as a source of exemplars for both the practice of scientific communities and for the public rhetoric of science. In this introduction, we discuss some questions about the nature of scientific biography in relation to literary biography, the historiography of science, and the public representation of the scientists, prompted by our reading of the various chapters in the collection.

Self and science

Science, by self-proclamation and popular repute, is objective knowledge. This status distinguishes it from other kinds of knowledge or belief, such as philosophy, literature and sociology. Conventionally, science proceeds by the application of a rigorous methodology, designed to eliminate personal bias through a variety of procedures such as replication, controlled studies, peer review and so on. Whether science is, indeed, objective knowledge, whether it proceeds by the application of a singular method, and whether the relation of the producer of knowledge to that knowledge distinguishes science from other knowledges are questions that have been the subject of considerable recent debate, spilling over into heated disputes in the pages of newspapers and popular books (Broderick 1994; Wolpert 1992; Fuller 1994; Gross and Levitt 1994).

A striking piece of evidence for the disinterested, objective status of science is the peer-refereed scientific paper, which in its structure, substance, language and idiom suggests that science is a rational, unambiguous and impersonal route to the acquisition of knowledge. That this suggestion is over-simplistic, erroneous, even fraudulent, was the remarkable claim of Peter Medawar a generation ago (Medawar 1964). Medawar showed the fragility of many of the claims for the special status of scientific knowledge; and his paper has often been cited in the recently burgeoning field of studies examining the ways in which scientific prose seeks to persuade readers that science is a mode of truth whose claims faithfully represent the order of nature (see for examples Gross 1990 and 1991). In a study of scientific writing, David Locke says that 'science has made a shibboleth of denying the expressivity of its own discourse', while Gyorgy Markus explores the way in which the 'inscribed author' of the scientific article appears as an 'anonymous performer of methodologically certified, strictly regulated activities and as a detached observer of their results — without any further personal identifying marks beyond possession of the required professional competence' (Locke 1992: 59; Markus 1987: 13).

Those who do not practise science encounter its claims to objective status principally through non-scientific media: listening to a scientific expert introduced on radio or being guided across the universe by a science broadcast on television. Sometimes, but more rarely, they may find scientific expertise deployed once again as objective knowledge, in a law court or a doctor's surgery. Few lay people have direct access either to scientists or to their technical writings, but for over two centuries they have had easy access to biographies and autobiographies of scientists. These offer a rare opportunity for lay people, uninvited to the laboratories and conferences, and effectively barred from access to professional journals, to listen to scientists in their own words. How do these scientists speak, and appear?

The question has been infrequently asked, but it is worth asking since it provides us access to some of the particular features of scientific biography, a genre which has several features in common with autobiography. In 1968, the chemist Erwin Chargaff wrote of scientific autobiography as belonging to a 'most awkward literary genre', written by people who typically lead 'uneventful lives' and offering 'the account of a career, not of a life'. Chargaff added that the career is likely to lack personal interest because, contrary to the situation in the arts, 'it is not the men that make science; it is the science that makes men' (Shortland 1988: 172). This succinctly expresses science's proclaimed objective status, in a form that refers explicitly to autobiography, while at the same time suggesting that scientific biography, too, is likely to 'lack personal interest'. The point can be rendered in any number of ways: scientists are typically dull, but science is exciting; the life of a scientist is a life of public acts, not private, still less intimate, details; without Milton there would have been no 'Paradise Lost', but the universal law of gravitation would have been discovered by another if Isaac Newton had never been born. It was the essayist and journalist Walter Bagehot, author of an improbably titled classic *Physics and Politics* (1872), who gave one of the best expressions of the scientist's irredeemable boringness. Prefaced by the remark that some people are 'unfortunately born scientific', Bagehot wrote in 1856 that those with an interest in 'shells, snails, horses, butterflies', who have 'delighted in an ichthyosaurus, and excited at a polyp', who are 'learned in minerals, vegetables, animals', are the consequence of an 'absence' within their constitutions of 'an intense and vivid Nature'. Scientific men are by nature dull and frigid and calm. '[A]n aloofness and abstractedness cleave to their greatness. There is a coldness in their fame, [whereas] the taste of most persons is quite opposite' (Bagehot 1965–86, 1: 397–98).

Such ideas *marginalise* the creative involvement of individual scientists in the production of knowledge. They underscore several well-known scientific autobiographies and present us with something of a paradox: that scientific autobiographies seem to present accounts of the role of the individual at odds with that offered in scientific biographies, indeed, with those presented in the conventional discourse of science which celebrates individual achievement. A few examples of scientific autobiography may serve to introduce the paradox.

Einstein's *Autobiographical Notes*, published in 1949 when he was 70, came into being through the unrelenting persuasion of an editor, and presented a formal, 'intellectual' portrait. In a short piece entitled 'Self-Portrait', written thirteen years earlier, Einstein had explicitly declared his lack of interest in the emotional tangles of self-knowledge: 'Of what is significant in one's existence one is hardly aware, and it certainly should not bother the other fellow. What does a fish know about the water in which he swims all his life?' (Einstein 1949: 5). Not surprisingly, Einstein's autobiography consists almost wholly of physics and with the exception of an occasional sentence, all but the first six or seven pages are inaccessible to the non-scientist. Here, though, reading carefully and piecing together some fragmentary clues, we can grasp what Einstein calls 'a definite form' to his life and its promptings not only 'from the outside' (1949: 95, 5). Einstein speaks of early religious yearnings, for example, linking these to his attempt to free himself from the 'chains of the "merely-personal", from an existence dominated by wishes, hopes and primitive feelings' (1949: 17), but always returns to a benchmark definition of himself as a man of intellect, not emotion or experience: 'The essential being of a man of my type lies precisely in what he thinks and how he thinks, not in what he does or suffers' (1949: 17).

If Einstein's autobiography suggests a transcending of self, Freud's shows that he struggled with issues of selfhood throughout his life. Einstein situates his character in the public realm, finding fulfilment in the realm of thinking outside the traditional literary personality. Freud's *An Autobiographical Study* offers a comprehensive history of a man's work, thought processes, and ideas, first published in 1925 (Freud 1959). The text is masterly: succinct, accessible, compelling. But any reader of modern biography will be struck by what Freud leaves out of this, in Porter's terms, 'deed-oriented deadpan autobiography' (p. 215). In the 1935 postscript to the second American edition, Freud comments on the relation between life and career in his autobiography:

> Two themes run through these pages: the story of my life and the history of psycho-analysis. They are intimately interwoven. This Autobiographical Study shows how psycho-analysis came to be the whole content of my life and rightly assumes that no personal experiences of mine are of any interest in comparison to my relations with that science.   (Freud 1959: 71)

The 'story of [Freud's] life' that is 'intimately interwoven' with the history of psychoanalysis is the account of his professional life, for there is no personal life to speak of in the autobiography. What then is intended by the statement that psychoanalysis 'came to be the whole content of my life'? The suggestion appears to be that Freud's whole life was his professional life, but if this is the case, the reference to 'personal experiences', makes no sense at all. The threads of truth, self-justification, self-erasure and evasiveness knotted together here are difficult to unravel. Freud might be suggesting bewilderment with the boundaries of the public and private or, just as likely, his rejection of the personal life. But while this may be part of the style of science, it is hardly what we might anticipate from Freud (see also Young-Bruehl 1993).

Our last example of scientific autobiography, Charles Darwin's (1974), is probably the best known. This is as surprising a document, in its own way, as the memoirs of Freud and Einstein. Composed towards the end of his life (and first printed in 1887) when he could have looked back over honour heaped upon honour and a world-wide reputation, Darwin's autobiography does everything to minimise his contribution to the process and progress of science. Darwin tells us, for one thing, that he was never very intelligent or far-seeing – mediocre at school and university, generally dull. He informs us how he gradually lost all aesthetic sense, all enjoyment of poetry and literature, how he became 'like a man who is colour blind'. In fact – and here is the organising device that Darwin uses to construct his life story – he presents himself as a humble collector of facts (1974: 24). He approaches the world of nature (perhaps, we suppose, his life also) with the emotional blankness of a collector of evidence devoid of expectation, prior knowledge or subjectivity. Little wit (bad at school), not much imagination (can't fathom modern poetry), but lots of capacity to collect. 'I collected facts', he writes, 'without theory on a wholesale scale', adding that, 'my mind has become a kind of machine for grinding general laws out of a large collection of facts' (1974: 68).

Darwin, indeed, adopts the ultimate defence against any charge of authorial subjectivity: the defence *ex morte*. Having cast self from science, he casts it even from his life story: 'I have attempted to write the following account of myself as if I *were a dead man* in another world looking back at my own life' (1974: 8). A morbid impulse? Or a scientific one? There are certainly historical, social and, indeed, personal contexts determining what presentation of self such scientists as Darwin, Freud and Einstein are able to make. Darwin seeks approbation not for himself but for his own work or rather for the results he has introduced into science (see Gagnier 1991: 258–65). Freud's motivation in erasing his own presence from his autobiography seems to have been similar to Darwin's. A quarter of a century later, Einstein explains the pursuit of science as the application of a method which transcends the 'merely' personal. If such a life story lacks personal interest, this is attributable not to the author's lack of interest but to the values in place in the profession. A scientific autobiography is indeed 'a most awkward literary genre'.

Of course, this does not mean that it has not been a favoured form among men of science. Robert Boyle, Isaac Newton, John Flamsteed, and John Wallis – to name just English natural philosophers – all left autobiographies, or at least fragments of these; and this reflects the fact they were intensely introspective, concerned with their motives and missions (for examples, see Delany 1969). As seen in the more recent examples above, autobiographies continued to be written by scientists into the twentieth century, although as we have just noted, they seemed to tend towards the type of *homo clausus* – the contained rational (male) personality guarding against the intrusion of emotional, somatic influences (see Outram 1989). Or, as Regenia Gagnier has remarked of some by nineteenth-century men of science such as Darwin and Huxley, they 'identify themselves as objective scientists by differentiating "science" from subjective experience' (Gagnier 1991: 258; compare Cockshut 1984).

Two chapters in this volume deal with autobiography and its links with scientific biography. Dorinda Outram discusses the autobiographies of French savants at the time of the French Revolution. In doing so, she makes the important methodological point that these writings should not be treated as 'private documents', or as raw, primary sources for biographical studies of the individuals who wrote them. Rather, they were public statements written at a time of both historical and personal crisis, as the trauma of the Revolution upset earlier conditions of patronage and notions of scientific vocation. Outram reads these autobiographies – by men such as Antoine Lavoisier, Antoine Fourcroy, Georges Cuvier and Bernard de Lacépède – as attempts to negotiate the changed situation of the scientific elite who now found themselves as holders of technical expertise in a regime that looked to 'Nature' and 'Reason' as grounds of political legitimation. The autobiographical form allowed them to clarify a sense of vocation under these new conditions – one that combined the Stoic ethos of calm self-possession and the idea of Nature as an emotional resource. Her chapter thus shows that although these autobiographies should not be taken as a simple access to the life of the individual scientist, they tell us much about the emerging authority of the scientific community. In this light it is interesting that these late eighteenth-century autobiographies (in contrast with nineteenth-century ones) did not connect such authority to great moments of discovery, but rather to the idea of a choice about a certain kind of life path.

David Knight's chapter on Humphry Davy concerns a man of science who made exhilarating discoveries early in his life and lived long enough to see these as a distant glimmer. Unlike the savants considered by Outram, Davy did not publish autobiographical material until near the end of his life. In these writings, he reflected on the meaning of his scientific successes in a way that did not make them a simple preparation for philosophical and moral insights, thus complicating one of the standard apologies for science – that it allowed a movement from physical to moral knowledge. Knight considers the problem of writing a biography (as he has done) of a subject who has also written an autobiography, especially one such as Davy's which ponders the passage from science to wisdom.

Thomas Söderqvist's contribution to this collection summarises the development of scientific biography (he prefers the term 'science biography') and also offers a compelling and fresh perspective on the Nature and role of the self in science and in the biographical narratives of scientists. Noting the disdain of some scientists for the 'merely personal' aspects of their lives, Söderqvist finds that recent developments in the history and sociology of science show a similar dismissal of the authentic and central role of individual passion, personality and achievement in science. Recent works in the history and sociological studies of science have accented the social and rhetorical construction of science, and several deny the existence of any constraints on the operations of political, linguistic, or other factors in shaping scientific knowledge. In response, some contend that nature constrains what can be claimed as scientific knowledge, while others locate such constraints in the routines of experiment, the nature of language, even in the wiring or operations of the brain. Söderqvist takes a different tack across these choppy waters. The constraint that concerns him relates to the existential conditions of the scientist, conditions irreducible to the particular conditions of socialisation or acculturation, to class, rank or position in a scientific field, or to cognitive apparatus. For what remains *after* the sociological tide has receded is the freely acting, responsible, asserting individual scientist. Söderqvist's mission is to turn our attention to this much-neglected figure, not as a means of getting us back into sociological waters — biography is not, for him, as it appears to be for such of its recent defenders as Hankins (1979) and Robert Young (1987 and 1988), yet another means of documenting the social factors constituting science — but instead a means of grounding the production of scientific knowledge in the success or otherwise of each scientist's struggle for self-assertion. Where many see science as the key site for the operation of the paradigm, the discipline, the text, Söderqvist finds recoverable there, via careful biographical work, evidence for the individual's struggle for existential authenticity in the face of socio-political constraints. The 'ability to handle the enabling conditions of self-assertion', he writes, 'lies at the heart of the life and work of every scientist' (p. 66).

One obvious objection to Söderqvist's work is that it seems to return us to a vicious psychological essentialism. Countering this impulse is a wealth of new interpretative accounts, ranging from Frank Sulloway's work on birth order (Sulloway, in preparation), through Michael Sokal's contribution to life-span developmental psychology (Sokal 1990), to revisions of psychoanalytical theory (Shore 1980; Hoffman 1984; Flanagan 1982). Each of these approaches could become reductionist: as Freud said of psychoanalysis, 'If anyone holds out a little finger to it, it quickly grasps the whole hand' (Enright 1989: 349). Yet, Söderqvist's recipe is for open biography and, as a recent study of Jean Piaget indebted to Söderqvist makes plain the recipe can lead to several different dishes (Vidal 1994). Söderqvist ends his chapter with a call for scientific biographers to take risks, to experiment with narrative, explanatory structures and language, using 'collage, narrative discontinuity, multigenre narratives, unsuspected time-shifts . . . stream of consciousness, symbolism, poetical reconstructions, and polyvocal texts'

(p. 79). If this sounds like incoherence, it would be a very attractive variety of incoherence and one which biography all too seldom achieves. It would also avoid what James Clifford has called the 'myth of personal coherence' (Clifford 1978: 52).

The myth of 'personal coherence' has shaped and stunted biography. Many have claimed that the sense of personal coherence, deriving from an ordered narrative, arranged interpretatively or chronologically, and the expression of a subject's inner life (what Frank Vandiver terms the 'essence of a subject' [Veninga 1983: ix]), is 'elemental' to biography (Nadel 1984: 155). Thus biography becomes a means of evoking an essential character and personality, an agent, in effect, of humanism (see Veninga 1983: 4–5). The self becomes a precious object and possession, something celebrated like a totem, as Claude Lévi-Strauss remarked (Clifford 1978: 49); order and coherence are paramount. Questioning the myth makes it possible to produce a 'narrative of transindividual occasions', weaving the collective through the individual (Clifford 1978: 52). Individuals become meeting points for influences, no longer static but mobile, effusive, decentred, a process not a thing. Söderqvist's pioneering contribution sets out to find a means of developing the narrative Clifford proposes without losing touch with the authenticity of the free, creating, self-motivated subject. Perhaps, as Söderqvist seems to suggest, some clues may lie in fiction, in such novels as *War and Peace* and *Middlemarch* which toss individuals into broad historical processes while retaining for and in them the traces of individuality, and without finding in them some essential character or possessed personality; or *Remembrance of Things Past* whose violations of chronology offer precious truths about character (Edel 1981: 10; 1978: 145).

Periphery and centre

Turning now to a broader picture of the discourse of science, in its historical context, we find that the celebration of the work of individual practitioners has been fundamental to the practice of science, or, at least, to the rhetoric of that practice. Although biography was marginalised by most of the influential historiography from the 1940s, biography and history have long coexisted as approaches to the study and promotion of science. However, attention to historical context reveals subtle, and dramatic, shifts in the role allotted to the individual in the creation of scientific knowledge.

The celebration of individual achievement was fundamental to the apologetics of science as it emerged as a distinctive practice in the early modern period. Even Francis Bacon, whose inductive method was supposedly aimed at levelling wits and understandings, allowed a place in his *New Atlantis* (1627) for statues to 'all principal inventors' (Bacon 1974: 246). As later commentators such as Whewell observed, Bacon was restricted in his opportunities for such celebration because he stood at the beginning of the revolution in scientific advance. But from about 1700, the practice of delivering formal éloges marking the lives and achievements of leading French savants was established by the Académie des Sciences. As secretary of the Académie from 1697, Bernard Le Bovier de Fontenelle used these occasions to engage in the controversy about the relative merits of the Ancients and Moderns, arguing that it was in science (not literature or art) that his contemporaries had surpassed earlier achievements. Furthermore, he drew upon the long-established discourse on the link between virtue and knowledge to claim that the moral character and behaviour of natural philosophers sustained their striking intellectual attainments (Paul 1980; Outram 1978; Weisz 1988).

Given this acknowledgment of individual achievement, what, then, was the attitude to biography in the period from the Scientific Revolution until the start of the nineteenth century? To answer this, we should clarify the relationship between biographical and more general historical accounts of science in the early modern period. This is a complex issue, partly due to the fact that, before the eighteenth century, histories of the mechanical arts and sciences were conceived in the Baconian sense of natural histories or catalogues of events. In his *Advancement of Learning* (1605) Bacon called for 'a just story of learning, containing the antiquities and original of knowledge and their sects, their inventions, their traditions' and the 'causes and occasions' of these. Yet in Bacon's classification of knowledge, history included biography — as the chronicle of a life. He argued that 'lives, if they be well written', could offer a level of detail not possible in larger chronicles. Indeed, Bacon wondered why the 'writing of lives should be no more frequent' since there were 'many worthy personages that deserve better than dispersed report or barren elogies' (Bacon 1974: 68, 72, 75). This understanding of biography was explained by Thomas Stanley, in *The History of Philosophy; Containing the Lives, Opinions and Actions, and Discourses of the Philosophers of Every Sect* (1701). There were, said Stanley, 'two kinds of History; one represents general affairs of State; the other gives account of particular persons, whose lives have rendered them Eminent' (Stanley 1701: preface). Similarly, 'Historical Dictionaries' published at the close of the seventeenth century, such as those by Louis Moreri and Pierre Bayle, included biography, together with geography. Peter Burke has noted that it was common in the early modern period for histories of learning to depend on biography as their frame of reference. The exception was the fine arts: because very few biographical details were known about the ancient artists, this writing could be only a general history rather than a detailed study of early individuals. Thus the distinction between biographies of artists and more general histories of art appeared by the Renaissance (Burke 1991: 10–11).

Some early historical accounts of science were influenced by ancient biographical models, such as Diogenes Laertius' *Lives of the Philosophers*. The Renaissance mathematician, Bernardo Baldi, produced a collection of biographies of mathematicians from the Greeks to Copernicus, taking Laertius as his model. Written in 1586–7, excerpts of 365 of these biographies were published as *Chronica de' Mathematici* in 1707 (Laudan 1993: 23 n. 11). But in 1696, Daniel Le Clerc issued an early call for a more thematic history of medicine: '[t]here is a big difference between writing the history of biographies of physicians . . . and writing the history of medicine, studying the origin of that art, and looking at its progress from century to century and the changes in its systems and methods' (cited in Burke 1991: 14). This search for the proper form for recording and understanding the achievement of natural philosophers continued through the seventeenth century. Michael Hunter's chapter highlights this issue: the activities of a seventeenth-century natural philosopher, such as Robert Boyle, required an intellectual biography – before this genre emerged. The memoirs by Burnet and Birch focused on his moral character, and were in fact utilised in the entries of eighteenth-century biographical dictionaries which had 'character' as their organising device. On the other hand, the abortive work by William Wotton which aimed to contextualise Boyle's discoveries sought to do so without analysing the debates in which Boyle was involved. By the early eighteenth century Boyle's earlier dominant position was slightly lowered as his scientific achievements in chemistry began to be incorporated in general accounts of Newtonian philosophy (in encyclopaedias and other popular works), although his character as a Christian Philosopher continued to be important, especially as more evidence of Newton's heterodoxy began to appear. This chapter thus suggests that, even in the case of a major discoverer, scientific biography of the 'modern' Boswellian kind, was not attempted before the nineteenth century.

The absence of 'modern' scientific biography may also have been affected by the passion for more general, philosophical history favoured by Enlightenment thinkers. About sixty years after Le Clerc's call for a thematic approach to the history of medicine, Jean-Etienne Montucla's *History of the Mathematical Sciences* (1758) broke with a biographical focus and betrayed the growing influence of Voltaire's philosophical approach to cultural history. Criticising the humanists who denigrated mathematics, Montucla regarded its progress as an index of 'the history of the human mind' (Laudan 1993: 6). This epitomised the Enlightenment historiography in which science was conceived as a rational progress not dependent on particular individuals. David Hume outlined a less idealist version of this approach in 1742 in his essay 'Of the Rise and Progress of the Arts and Sciences', suggesting that the pursuit and advance of science depended on social and political factors, such as trade and type of government (Hume 1993: 56–76). The most direct statement of the view that the progress of science was not solely a product of individual genius appeared in the historical works of Joseph Priestley. In 1767 he published *The History and Present State of Electricity* and followed this in 1772 with *The History of the Present State of Discoveries Relating to Vision, Light, and Colours*. In the first of these, Priestley made several remarks aimed at deflating the role of genius – even that of Newton – and championed the role of factors such as correct method and the accumulation and communication of data. On this view, science was accessible, and indeed, needed the participation of a wide range of people. Nevertheless, the first volume of the history of electricity considered, in separate chapters, the experiments and discoveries of individuals from William Gilbert to Benjamin Franklin. Admittedly, Priestley's general point was made by the inclusion of contributions from some whose names were not among the famous natural philosophers, and by mentioning not just success, but failure, as a part of science (see McEvoy 1979; Schaffer 1986).

Attitudes such as Priestley's are often erected into a general Enlightenment preference for abstract laws of progress that overshadow any focus on the individual. But this is an exaggeration. Even in the *Encyclopédie*, with its rational classification of the sciences, Jean D'Alembert acknowledged in the Preliminary Discourse of 1751 that 'the history of the sciences is naturally bound up with that of the small number of great geniuses whose works have helped to spread enlightenment among men' (D'Alembert 1963: 69).

Moreover, this period saw the publication of massive biographical dictionaries and some attention, by the close of the eighteenth century, to the assumptions and methods of biographical writing. Although it is true that there were very few free-standing biographies on major scientific figures (see Stauffer 1930 and 1941), the lives of natural philosophers were certainly covered in dictionaries, éloges, and obituaries. Indeed, as Richard Yeo indicates in his discussion of biographical dictionaries in this volume, the English version of Bayle's *Critical Dictionary*, organised by Thomas Birch, courted the Fellows of the Royal Society and made a special point about its addition of entries on contemporary men of science. These were based on observations and memoirs by friends and critics, but also on the letters and autobiographical writings of some of the major scientific participants such as Kepler, Boyle, Flamsteed and Wallis.

In fact, even Priestley was not opposed to the notion of biography, or at least to a form of it, because two years before the history of electricity, he published *A Description or a Chart of Biography* (1765) which went through fifteen editions by 1820. Originally given as lectures at the Warrington Academy, this confirmed the message of the companion *Chart of History* (1769): that history, including the progress of natural philosophy, could be traced through the activities of great individuals who offered models of human reason and virtue. Priestley did, of course, stop short of hero worship, stressing that great improvements 'are not, therefore, in general, to be expected from men confined to their closets' (Priestley 1765: 2; see also Bazerman 1991).

By the early nineteenth century, these tensions in the Enlightenment approach to science were visible. In his Preliminary Dissertation (two parts, 1815 and 1824) for the *Supplement* to the *Encyclopaedia Britannica*, John Playfair attempted a history of the physical and mathematical sciences. On the one hand he stressed the importance of proper method and procedures as guarantors of intellectual progress, yet he still identified individuals as the moving spirits of the age. Playfair referred to developments in natural philosophy between 1663 and 1730 as belonging to 'the period of Newton and Leibniz', after 'the men who impressed on it its peculiar character' (Playfair 1815–24, 4: 1). However, he recognised that there was another perspective that gave a less crucial role to the individual. Speaking of Newton he remarked: 'though the creative power of genius was never more clearly evinced than in the discoveries of this great philosopher, yet the influence of circumstances, always extensive and irresistible in human affairs, can readily be traced' (Playfair 1815–24, 4: 79). Among these he mentioned the state of astronomical knowledge at the time and the fact that at least one source of error – the Cartesian vortices – had been removed.

These examples suggest that historical explanations of the rise of science had begun to move biography from the centre. By the 1830s, a more definite break with biography as an organising principle was discernible in the two major historical panoramas of science by Comte and William Whewell. Comte's *Course of Positive Philosophy* (1830) offered a three-stage history of human thought – the theological and metaphysical stages giving way to the 'positive', exemplified by the method and concepts of some of the advanced physical sciences in the early nineteenth century. But in spite of this abstract analysis, Comte was willing to celebrate certain individuals as carriers of the intellectual spirit of their day. Frederic Harrison, one of his English disciples, edited *The New Calendar of Great Men: Biographies of the 558 Worthies of all Ages and Nations in the Positivist Calendar of Auguste Comte* (1892). Divided into thirteen months, this charted the development of human thought from its theocratic beginnings to the recent stage of modern science. Each month was named after a representative figure, with the French physiologist, Xavier Bichat as the 'saint' of the last month – modern science. Each month was divided into four weeks (astronomy, physics, chemistry and biology) with names attached to each of the seven days. Accompanying the calendar were snapshot biographies of these names which, in the case of the men of science, said little about them as individuals apart from the relevance of their work to the ineluctable march of positivist thinking.

Whewell's *History of the Inductive Sciences* (1837) was written in terms of the developments of the main branches of science. These were classified according to the Fundamental Idea (such as Space, Force, Final Cause) that set their particular conceptual framework. Like Comte, Whewell saw the progress of science occurring in three stages – a prelude, inductive epoch and sequel – but these took place within each of the different scientific areas, such as the physical, biological or historical sciences. This perspective could be regarded as one that interpreted the process of discovery as dependent on conditions, such as the clarification of the Fundamental Idea and the availability of empirical evidence that transcended individual contributions. Whewell still told the story through the activities of great discoverers such as Kepler, Newton and Faraday, although he abstracted and generalised their common intellectual characteristics to such an extent that he almost prescribed the qualities necessary in all future great discoverers: 'distinctness of intuition, tenacity and facility in tracing logical connection, fertility of invention, and a strong tendency to generalisation' (Whewell 1857, 2: 139).

The reaction to Whewell's *History* focused debate on the differences between biography and history as accounts of the nature of science, an issue latent in the earlier work of writers such as Priestley and Playfair. On one hand, some reviewers, such as John Herschel, saw Whewell's work as a useful corrective to the tendency of recent utilitarian reformers to reduce the importance of individual genius in order to promote the accessibility of science in their educational campaigns. In Herschel's view, these commentators incorrectly ascribed the progress of science wholly to correct method, or to the favourable conditions of 'the age'. On the other hand, David Brewster was not satisfied that Whewell's search for the Fundamental Ideas appreciated the irreducible genius of great discoverers, even though Whewell rejected any mechanical application of Baconian method as an account of major scientific advances. In his *Life of Isaac Newton* of 1831, Brewster explicitly contrasted history and biography as modes of grasping the essential personal qualities of great scientists. Anticipating the later dispute with Whewell, he explained that he did not dismiss the notion of distilling the 'general character' of scientific thinking, but asserted that 'the history of science does not furnish us with much information on this head, and if it is to be found at all, it must be gleaned from the biographies of eminent men' (cited in Yeo 1988, also Yeo 1993: 164–5). Taking the opposite view, Augustus De Morgan later said that in his larger work on Newton, Brewster was 'still too much of a biographer, and too little of an historian' (De Morgan 1855: 310). These comments suggest that the close ties between biography and history assumed in the Baconian classification were now broken or, at least, no longer unquestioned.

It would be wrong to see this debate on the relative merits of history and biography as a preview of the twentieth-century marginalisation of scientific biography, and for at least two reasons. First, the nineteenth-century discussion has to be seen in the context of the assertion of biography (including biography of scientists) as an independent genre. The publication of separate books on the biographies of men of science did not occur until the early 1800s, when writers and publishers exploited a new market for 'improving' texts, and scholars began to research primary sources (Sheets-Pyenson 1990: 399). Thus Brewster's was the first biography of Newton in English apart from entries in biographical dictionaries, such as the one Jean Baptiste Biot contributed to the *Biographie universelle* in 1821 (see Theerman 1985; Yeo 1988).

Second, the negative attitude to biography from the 1940s was supported by assumptions about the priority of method, procedures, paradigms, or shared metaphysical concepts over particular intellectual or emotional factors in the lives of individual scientists. Thus, in his effort to delineate the norms of the scientific community, Merton said that the personality and moral integrity of individual members were not important (Merton 1973: 276). In contrast, however, a concern with the individual man of science, or natural philosopher, was crucial to the practice of science from the 1650s, even though there were also statements about the importance of Baconian method and theories about the political and institutional conditions affecting the progress of reason. Again, Priestley illustrates how the Enlightenment interest in generalisations about the nature of science coexisted with a concern about the qualities of its individual practitioners:

> A PHILOSOPHER ought to be something greater, and better than another man. The contemplation of the works of God should give a sublimity to his virtue, should expand his benevolence, extinguish every thing mean, base, and selfish in his Nature, give a dignity to all his sentiments, and teach him to aspire to the moral perfections of the great author of all things (Priestley 1775, 1: xxiii)

To some extent, Yeo's chapter offers a perspective on these relations between biographical and historical accounts of the individual scientist. He examines the place of scientific biography in the major dictionaries and encyclopaedias from the eighteenth century. One significant point here is the absence of all biographical entries in the leading encyclopaedias, such as Chambers' *Cyclopaedia* (1728) and the *Encyclopédie* (1751). These presented themselves as dictionaries of arts and sciences dealing in systems of ideas; they excluded biography as more appropriate to so-called 'historical' dictionaries which included biography, along with geography. Yeo explores the contrast between an emphasis on the 'character' of men of science in the historical dictionaries, and the more impersonal accounts offered in encyclopaedias, when they finally incorporated biographical entries. In the early nineteenth-century editions of the *Encyclopaedia Britannica* these focused on the relative contribution of individuals to the ineluctable progress of knowledge, rather than on the 'moral perfections' that Priestley urged natural philosophers to cultivate. However, this should not be taken as a simple contrast between the Enlightenment and the nineteenth century: by the 1830s, biographical accounts of scientists, in memoirs, periodical articles and popular books aimed to evince precisely how their subjects lived up to such high moral expectations.

Victorian biography: dead and buried?

If any historical period has displayed an interest in biography (including scientific biography), equal to our own, it must be the Victorian era. Indeed, the nineteenth century as a whole witnessed a massive efflorescence of writing in the first person singular: the maxim, essay, diary, notebook, the letter, the self-portrait, the autobiography, the memoir, the first-person novel, the lyric — the entire range, in short, of what French structuralists used to call unflatteringly *la littérature personelle*. From 1813, when a denunciation was published of the contemporary 'rage for indiscriminate biographical reading', to 1884 when a critic boasted that England had published more biographies than any other country, the period saw an enormous number and enormous variety of biographical studies (Stanfield 1813: 335; Christie 1884: 187–88). Some critics, like Thomas Carlyle writing in 1832, attacked the frivolity of much biography and the popular interest it encouraged in 'Gossip, Egotism, Personal Narrative . . . Scandal, Raillery, Slander, and such like' (Carlyle 1832: 253–4); others denounced biography for its elevated pretensions and bulk. 'Scarcely any man of note can get safely out of this world without leaving behind him . . . a son, or a friend, or a professional man of letters, ready to "take him off" and set forth his portrait in black and white, in voluminous volumes' (Anon 1879*b*: 255).

'[I]ll-digested masses of material . . . slipshod style . . . tone of tedious panegyric . . . lamentable lack of selection, of detachment, of design . . .'. Few can have approached nineteenth-century biography without denunciations such as these ringing in their ears. They are drawn from the preface to Lytton Strachey's *Eminent Victorians* (1918), a book that manages the feat of denouncing and yet promoting biography, indeed, burnishing it (in a frequently quoted tag) as the 'most delicate and humane of all the branches of the art of writing' (Hamilton 1994: 8). Strachey's book is intended to attack one style of biography and to provide a manifesto for, and exemplars of, another. With its bold, paradigm-shifting preface, calling for biography to be shaped by the values of honesty, brevity and art, *Eminent Victorians* has been taken as laying the basis for modern biography, and Strachey's caustic comments on his predecessors have often been repeated (see, for examples,

Veninga 1983: 13; Gittings 1978: 37; Hoberman 1987: 5). A typical recent statement finds that '[t]he nineteenth century did little to advance the cause of life writing' and that we must turn to the last sixty years or so for interesting developments (Kingsbury 1987: 230).

The predominant image of Victorian biography, owing largely to Strachey, is of concealment and 'sugar coating' (Cox 1987: 370): '[a]n age that invented the verb "to Bowdlerize" and considered "leg" an indelicate word could not be expected to excel in biography' (Garraty 1957: 94). These are lives in which neither illegitimate children nor adultery, neither mental anguish nor sexual peccadillo exists. We learn nothing from John Foster's 1872–74 biography of Dickens' bouts of dark depression nor of his mistress Ellen Ternan; nothing from Mrs Gaskell's 1857 *Life of Charlotte Brontë* of her subject's consuming passion for Monsieur Heger. The situation is little different in scientific biographies, particularly those written under the watchful eye of a scientist's widow or other relative. One example may stand for many.

Katherine Lyell, Sir Charles Lyell's sister-in-law, brought out the *Life, Letters and Journals of Sir Charles Lyell* in two stout volumes in 1881, six years after the geologist's death. These have become a standard source of information on this major scientific figure but the text contains omissions, ranging from single key words to an entire significant letter, which burnish and preserve a favourable image of Lyell. For example, in a letter to Gideon Mantell, a reference to John Murray as a 'tradesman', a comment that 'Featherstoneaugh has made an ass of himself by a poem on the deluge which is despicable [sic] low and vulgar', and a reference to coprolites being 'parts of animals' are all deleted (Lyell to Mantell 23 April 1830). Other subjects which are censored or amended concern Lyell's occasionally hostile views of the Established Church, and his many references to income and money, so that his shrewdness and awareness of finances are minimalised. Some deletions clearly conform to early nineteenth-century upper class conventions, but others are more curious. For example, the name of George Toulmin, who had speculated about the indefinite antiquity of the earth, was suppressed, thus preventing readers knowing that Lyell had read Toulmin's work. While a complete account of these editorial intentions is not yet available, Katherine seems to have taken considerable care to prevent Lyell's career aspirations, and his intellectual and other debts, from being portrayed. The result is not only an incomplete portrait of Lyell, but one plainly at variance with the historical record (for contemporary views of the ethical issues, see Oliphant 1883; Purcell 1896; on the making and use of such works, see Browne 1978 and 1981).

Historians have pursued the contradictions at the heart of Victorian social and personal life – squalor alongside prosperity, anti-industrial values alongside commercial society, ideology of self-help alongside programmes for social improvement, aristocratic power and privilege that thwarted democracy, economic anxiety alongside expansive empire building. No one reading Strachey's portraits of Cardinal Manning (now the archetypal hypocrite), Florence Nightingale (bed-ridden tyrant), General Gordon (egoistical colonialist and religious megalomaniac) and Thomas Arnold (earnest Victorian) can see them as in any way eminent again. Following Strachey's pitiless candour, their reputations have lain in tatters. Even the greatest have fallen: Queen Victoria, who escaped Strachey's strictures (his biography of her is little short of adulatory), has been exposed in Weintraub's intimate biography (1987) as an unhappy, plain, overweight depressive, infatuated with Lord Melbourne, a woman who hated pregnancy, children and babies, and displayed little warmth for her own offspring. Indeed, Weintraub's study, with its evasion of anything resembling historical specificity, reinforces the Stracheyite assumption that the 'Victorian era', with its 'Victorian people' and 'Victorian mind', is an unproblematic, seamless web. Once again, the gap between modern historiography and biography is clearly visible.

Strachey's objection was in part that the Victorians left so much out of their biographies and in part that they put so much in. They knew nothing, he wrote, of 'a becoming brevity' that excludes everything redundant and nothing significant (quoted in Strouse 1978: 123). As Leslie Stephen put it: 'The reader should ask for more and should not get it' (Stephen 1893: 181). Brevity – Strachey's lives run to less than 100 pages each – compels the biographer to select and organise material carefully, to compose a narrative, to produce an artistic and coherent unity.

Strachey's targets are well selected. Who nowadays reads J.G. Lockhart's ten-volume life of Sir Walter Scott (1837–38), a once highly esteemed study? How many historians of nineteenth-century science are familiar with such biographies as the 'Life and Letters' of Hugh Miller (Bayne 1871), Richard Owen (Owen 1894), and William Whewell (Todhunter 1876)? On the other hand, some aspects of Strachey's critique are applicable to biographical practice today. Immediately after Strachey's book, short and satisfying biographies appeared by such writers as Virginia Woolf, David Cecil, Harold

Nicolson and Cecil Woodham-Smith, then the 500-plus page monster crept back (Gittings 1978: 64–5). What Gore Vidal dubs the 'trash-basket approach' to biography, so evident in the monumental size of nineteenth-century biographies, seems to be with us again today (Veninga 1987: 97). In scientific biography, there has been a spate of comprehensive books that seem to tell everything, perhaps more than one requires, without always defining what is essential to form a rounded picture of the subject. John Bowlby's biography of Darwin runs to over 500 pages but tells us little about the social and political contexts of his science; Desmond and Moore's biography of Darwin is over 800 pages long but devotes surprisingly little attention to the science at the heart of Darwin's major work. Both are splendid studies and each is fathoms away from the 'Life and Letters' approach of Darwin's earlier hagiographers. But the reader may wonder whether Nabokov's famous adjuration to his literature class to 'fondle the details' has not been taken a step too far.

The stop-go, rag-bag biography, another Victorian ghost, has made a reappearance. Nowadays biographies often swell with incorporated diaries, letters and suchlike into a narrative of the complete life and the almost-complete work, the whole organised, not (as were Strachey's) thematically but chronologically, like nineteenth-century biographies. From family ancestry, we come to the birth of the subject, followed by a brief youth and adolescence, before the biographer arrives at the first notable work. Here we stop for a description and assessment, copiously illustrated by testimony, before we move again. The subject lives through another year, another lustrum then another pause for breath, before we move again. (The style is typical of biographies by Lockhart or Forster or Buckle. A modern version is Irvin Ehrenpreis's life of Swift (1962–83).) Volume one (of three), for example, records Swift's early life, offers chapter-size slabs of criticism, one for the *Tale of a Tub*, another for the *Battle of the Books*, a third for the *Mechanical Operation of the Spirit*, then back to Swift's career, and a forward chronological motion. The same organisation occurs in other 'definitive' biographies, such as Edgar Johnson's two-volume *Charles Dickens* (1952) and Michael Holroyd's *Lytton Strachey* (1967–68).

An example of the stop-go approach in scientific biography, prompted perhaps by the particular difficulties of integrating technical material with a life, may suggest a radical separation between the life and the work. Crosbie Smith and Norton Wise's 800-page biography of William Thomson, Lord Kelvin (1989), gives a fascinating but compressed account of its subject's forlorn courtship, multiple rejections and rebound betrothal to another woman. This marriage is dealt with in less than a page, before the authors plunge back into mathematical analysis. Was Lord Kelvin able to separate his public from his private life? This would be remarkable, and noteworthy. Did his private life have no bearing on his physics? Again, this would be worth exploring. Maybe Lord Kelvin did not *have* a private life in any conventional sense. These are perplexing questions, and it is unfortunate not to have them raised or settled in this otherwise magisterial study.

The alternative to a radically bifurcated biography is the biography that strives for a narrative unity — without doubt the more common approach. Virginia Woolf saw the method of this biography as being to 'weld . . . into one seamless whole', the 'granite-like' solidity of truth and the 'rainbow-like infrangibility' of personality (Mandell 1991: 3). Such an approach makes possible a more coherent life story. And this coherence, in turn, offers a satisfying, even nostalgic, recovery of sense. Consider the conditions when the popularity of biography rises. In the 1930s, for example, there was a surge of public interest in biography in the United States. Contemporary commentators, especially emigrés from Europe, speculated about its cause. Leo Lowenthal, for example, charted biographies serialised in magazines and newspapers, emerging from Hollywood studios (the 'bio-pic'), and produced in bulk by the publishers of cheap, paperback literature. His view was that biographies fulfilled a need during a time of economic depression. Popular biographies offered inspiring tales of people who rose against the odds, people standing fast against conventions and outmoded traditions. 'We are exalted', wrote a commentator 'not by the lives of the saints but by the heroic exploits against great odds and unfeeling bureaucracies' (Veninga 1983: 68; and for a psychoanalytic account of this appeal, see Horney 1937).

Biography shapes the life of the subject from incoherent fragments, and it helps to shape the life of the reader. It expresses a kind of nostalgia for other people's lives. Albert Camus put his finger on this: when 'seen from the outside' the lives of others appear to 'form a whole', whereas our own life, seen from inside, 'is all bits and pieces'. And so we 'run after the illusion of unity' that is offered to us in biography (Clifford 1978: 17). On this view, when the experience of individuality is difficult to uphold, biographies offer hope, weaving narratives of centredness and subject power, tales in which individuals exert control over themselves and their surroundings. The achievement of unity is the end of biography, and the goal reached is tribute to a 'good' biography, one 'true to life' and 'believable'. Biography's distance from critical theory has insulated it from the dethroning of the confiding subject. While biography purports to present the 'real Me', criticism, since author and subject were pronounced dead by Roland Barthes (1977), has busied itself only with texts, readings and rereadings. This exposes once again the tremendous gulf that separates conventional biography and conventional criticism.

Passion and depth

Strachey's attack on Victorian biography has been by and large accepted. But historians, especially historians of science who so often have recourse to Victorian 'Life and Letters' of Darwin, Lyell, Hugh Miller, William Buckland, Adam Sedgwick and others, need to ask whether Strachey's dismissal is accurate. Did nineteenth-century biography eschew criticism of its subject? Did it ignore the private, intimate life in favour of the public act and published writing? Was it prim and prudish? Why, indeed, did Victorians produce biographies in such profusion?

From Strachey's time onwards, there has been a tendency to categorise in very simple terms the rich variety of biography. A good deal of what passes for biographical criticism and history amounts to little more than classification along structural, stylistic, rhetorical, thematic and other dimensions: the 'strippers' and the 'hiders'; biography that leans towards history or fiction (Kaplan 1978: 5; Butler 1967: 3); biography as the history of an individual, the exemplary life, the accumulation of anecdote (Nadel 1984: 104); and so on. Behind these schemes lies a distinction, almost matching that between the Aristotelian and the Platonic traditions, between biography that is logical and factual, written with a cool eye, and biography that is full of allusion and literariness, myth and hearsay, as well as evidence.

But this will not do, even in parody, for it prevents our seeing the full range of the nineteenth-century legacy in all its variety of style, aim, intended readership and substance. One of the merits of Martha Vicinus' contribution is to explore another side to Victorian biography, indeed, another kind of Victorian biography. Vicinus' focus is upon a little-studied set of biographies intended principally for a female readership. Produced in large numbers, popular lives of women offered a broad range of moral and didactic codes, along with ideals of public and private self-presentation and self-effacement.

Samuel Smiles' emphasis on determination, hard work and self-help looms large in several examples, never more so than in their purported aim to set standards of purpose, will and determination, coupled to codes of appropriate behaviour, that might, as Joseph Johnson wrote in 1863, 'incite his sisters . . . to become "Clever Girls" ' from the force of example (see p. 199). Reading Vicinus, it is difficult to suppose that anything as simple and emblematic as the 'Victorian biography' existed, for the negotiation of class, gender and personality, not to speak of the often ambiguous role of the narrator-biographer and the set role of the reader, shows biography to have been multivalent and heavily overdetermined. Vicinus' focus on Florence Nightingale reveals some unexpectedly complex aspects not so much to Nightingale herself as to her role in biographical and cultural narrative. Cast for several parts on different stages, Nightingale emerges as a figure of depth and ambiguity: to some a thoroughly modern girl, to others an old-fashioned heroine; here a woman of selfless public service, there a heroine selflessly battling against adversity.

On the whole, nineteenth-century biography was imbued with a kind of moral passion. The fundamental motive for writing biography in this period, as Ruth Hoberman has shown, was to describe an admirable role model, admirable because crowned with achievement (Hoberman 1987: 191). This explains the biographer's reliance on the documentation of public beneficial acts. If so, it would seem important to discover how biographies were read, and whether they were used as models. Recent work has suggested the importance of autobiography in 'self-fashioning', by providing scientists with exemplary models of selfhood and success (Shortland 1988, 1996) that become, in the words of the French geneticist François Jacob, 'the statue within' (Jacob 1988). In his contribution to this book, Roy Porter considers the presence of an imagined self in the writing of Thomas Beddoes, a late eighteenth-century physician who wrote profusely not only in such genres as the medical case history and the literary essay but also texts whose substance and style is less classifiable.

Although concerned with the relations between biography and illness, Porter's contribution is not a study of individual illness, using retrospective medical or psychiatric diagnoses. Such studies, for example of Darwin (Colp 1977, 1987; Bowlby 1990), have been entertaining but, in the end, fruitless searches for the *real* cause of Darwin's ill-health, based on the questionable assumption that the key to the development of Darwin's theory of evolution is to be found in his sickness (see Wilson 1977; Goldstein 1989; Smith 1990,

1992). One is reminded of similar once-bruited claims that the outcome of the Second World War hinged on Churchill's sleeping habits, the reign of Henry VIII turned on his gout, and Napoleon's inconsistencies were the consequence of arsenic poisoning. Even the great Jules Michelet once fell prey to this virus of pathography, when he expressed the view that the reign of Louis XIV could be divided into two distinct phases: before and after the king had developed a fistula (Sokoloff 1938: vi) – it was a sober judgement that has not received the acclaim of more modern historians! More recently historians have used the testimony of supposedly sick figures as entry points to understanding the medical world of the time: Roy Porter has considered Scottish attitudes to mental illness via an examination of the life and death (through suicide) of the geologist Hugh Miller (Porter 1996), John Wiltshire has thrown interesting light on the mid-eighteenth-century medical world by attending to Samuel Johnson (1991), as do Nora Crook and Derek Guiton in their exploration of Shelley and syphilis (1986). Porter's contribution to the present book takes a different approach.

On the face of it, in examining the past, present and future of people's bodies, doctors are professionally engaged in writing biography. There are, indeed, many affinities between writing and doctoring, and dealing with sickness is to engage with self-experience in a way normally closed off from daily routine. To be ill is to be made aware of one's own body, and often one's mind, in relation to others, and also of new social relations, to doctors, nurses and other patients. As sick man and a healer of sick men, Beddoes seems to have known plenty about the 'sick role', and his self-reflexive writings, with their queer digressions, expected banalities and arresting insights, offer something that is at once scientific, biographical and autobiographical. In his writing Beddoes acts the role of sage, critic, buffoon, hero, prophet. As Porter puts it, he projects himself in these different guises, guises which (and one doesn't need Freud's help here) offer him 'oblique gratification, allowing him to strut as an exemplary figure of sorts' (p. 299). Beddoes presents his biographers with a very real challenge: parody, farce and irony are just three of the idioms evident in his singularly autobiographical writings. Any biographer approaching Beddoes as subject will need therefore to be closely attuned to questions of narrative style and voice in his writings.

That someone as important in his time as Thomas Beddoes should have been only very recently accorded a biography that added anything at all to the sanitised account that appeared in 1811 is surprising; medical biographies, as Morton and Moore (1989) show, are hardly uncommon. The explanation lies in the vicissitudes of reputation. Reputations are made, not born, and they are subject to change. Beddoes' star fell substantially after his death, despite the efforts of his widow Anna to commission what Porter terms a 'highly doctored biography of her husband' and J.E. Stock's 1811 biography which downplayed Beddoes' extensive political activities. The fluctuating fortunes of a reputation, once traced and explored, can throw light on broader patterns of historical change, but this requires a sophisticated treatment of the relationship of text and context, subject and period. Ill-considered treatments find the reputation of a writer to be prey to the vagaries of 'taste', a term only marginally less obscure than that other curiosity, 'the verdict of posterity', as John Rodden has shown in his brilliant study of George Orwell (1989).

Turning to another Enlightenment figure, John Gascoigne's chapter reveals the subtle forces at work in the making and claiming of a historical figure such as Sir Joseph Banks. Gascoigne attends to the ingredients essential in the phenomenon of fame — international conditions that create and sustain a reputation, as well as more local, idiosyncratic factors. Posterity granted Joseph Banks (as it did Florence Nightingale) a mixed and unsettled standing, and it was never clear what would constitute the solid foundation of his repute, particularly since he did not bequeath any permanent record of publications. So how was Banks to be remembered? As scientist, traveller or patron; administrator, explorer or colonial advocate? Already in 1845 Henry Brougham noted how Banks' name 'could not easily find a place' in the annals of science, 'from the circumstance of its not being inscribed on any work or connected with any remarkable discovery' (p. 246). The problem created by Banks seemed to make any biographical study of him difficult since biography is so commonly organised and prompted by recognisable achievement or a corpus of work. This may be why his biography was so late in coming and why celebration and historical remembrance came about in the form of the recuperation of his papers, letters and other ephemera. Yet, biography is able to contain contradiction and ambivalence by imposing order through chronology or through centring on one determining achievement. Biographers turned Banks into a national icon, the 'father of Australia'. His life became not so much an exemplary individual achievement as a kind of conduit for the imposition on a barren landscape of civilising values. It was precisely Banks' lack of notable achievement that allowed him to be the vehicle for such forces as Enlightenment civility, progress and colonialism.

A good contrast with Banks is Michael Faraday, a scientific discoverer *par excellence*. Geoffrey Cantor, having recently written a biography of Faraday, uses his chapter to reveal the conflicting images of the man produced by Victorian biographers. Here again, we see that a restriction to the kind of literary evaluation of biography, offered by Strachey, misses the point. Cantor thus provides a comparative study of the great variety of biographical sketches and accounts of Faraday in periodical essays, obituaries, and in both popular and more scholarly books; he does not ask which of these provide the most authentic portrait of the 'real' Faraday, but rather analyses the way in which these writings constructed different images of the man of science. Given some of the negative contemporary references to the character of scientists, such as those mentioned by Bagehot, and what George Henry Lewes called the 'dread and dislike of science' (Lewes 1878), it is significant that these late-nineteenth century biographies drew upon exemplars that already enjoyed some cultural status. Cantor shows that Faraday was presented as either a romantic genius and seer, or a Smilesian hero who achieved fame through hard work and moral probity. To some extent, these two versions were associated with conceptions of science as either theoretically speculative or assiduously empirical.

These chapters (and also Knight's) therefore indicate the possibilities for using past biographies as sources in the study of changing constructions of scientific identities and public perceptions (see similar studies by Outram 1976; Yeo 1988; Theerman 1990; and Hunter in this volume). But this approach does not imply that, for those who write biography, the relationship between the biographer and his or her subject can be ignored.

Biographer and subjectivity

One of the earliest biographies, Plutarch's *Lives*, was written to provide the author with a 'sort of looking-glass, in which I may see how to adjust and adorn my own life' (Nadel 1984: 21). Yet, while the author not uncommonly appears in confessional mode in the introduction, he or she rarely explains why a particular biography was written and why in a chosen style. Stephen Marcus is an exception, describing in detail his motivations in tackling Dickens, and finding answers in his own youthful attraction to the novels, his conflict with his own father about the study of literature, and his being granted 'inner permission' to proceed (Baron and Pletsch 1985: 3–4, 298–9). This kind of auto-criticism will strike some as unnecessary and self-indulgent, and indeed, there are times when the author's voice seems intrusive. Yet self-analysis is a necessary process, if we are to be alert to our own (unconscious) idiosyncrasies as we write about those of others.

A little-known but fascinating paper by Richard Westfall, the biographer of Newton, offers indications of the value of introspective study of science to the historian (Westfall 1985). It deserves careful examination. Westfall realised after having completed his enormous biography, some twenty years after setting out, that aspects of his relationship to Newton had, perhaps, influenced his writing. At the start of his biography, Westfall offers the puzzle that 'the more I have studied him, the more Newton has receded from me' (Westfall 1985: x). Written at the end of his study, this seems to present a standard, Socratic, statement of ignorance ('the more I know, the more I know how little I know'). It also represents, less obviously, a kind of mourning for a subject now lost, who for most of the author's mature life has been a constant presence by his side. But what the reader of the preface and biography would not know is anything about how Westfall's own views might have influenced his portrayal. In a subsequent interview, when quizzed by an analytically inclined interviewer, Westfall explains that he is a Presbyterian elder, and emotionally as well as intellectually wedded to the values of his church. This casts interesting light on Westfall's investment in Newton and the commitments which shaped his understanding of his chosen subject.

Westfall is explicit now that the choice of Newton as a biographical subject was not, at the outset, the result of a sequence of purely academic interests (Baron and Pletsch 1985: 7). He is equally explicit that the problem he confronted was not simply the problem of dealing with any historical, receding, subject, but that of Newton's alchemical and theological activities. Westfall identified strong feelings and attitudes towards these, and these needed to be dampened for the biography to be successfully brought to a conclusion. The statement that 'the more I have studied him, the more Newton has receded from me' seems to confirm that Westfall has returned home unaccompanied, perhaps even unaffected, by Newton's heresies. Looking again at his portrait of Newton, Westfall acknowledges that he drew there 'an *ideal self*, with 'an ethical dimension' (Westfall 1985: 187, our emphasis):

> I presented Newton as a faithful steward, a man to whom his
> master had entrusted an extraordinary talent, a man determined
> not to bury his talent in the ground but to employ
> it in order that he might return more than he had received
> (Westfall 1985: 187–8)

But who *is* this 'ideal self'? It is Newton as he appears in Westfall's biography, and it is also Westfall himself. The point may seem obvious: in a recent review of Newton biographies, B.J.T. Dobbs remarks pithily that '[w]hat we already have in our psyches and our intellects we tend to find in Newton' (1994: 516). But it is not simply that Westfall's biography is autobiographical, it is, rather, that Westfall's is an *ideal* autobiography. Westfall's Newton is Westfall as he would like to have been. So in what is a key trope in the book – given in the title, *Never at Rest* – Westfall suggests that Newton's was a 'ceaseless search for truth'. This is striking when set against a social and cultural environment which was (at least at Cambridge University) in marked decline, and when the typical pattern for the age was inactivity. Newton could have been portrayed not as a ceaseless worker but, say, as a cloistered don, a monkish natural philosopher. The materials which appear as evidence in Westfall's biography are determined by an image of Newton's selfhood which is in turn determined by Westfall's own sense of what constitutes an ideal self both for his subject and for his own possession. Without denying that this is a valid portrayal, it is not the *only* one, and Westfall, guided by introspection, recognises that his choice of Newton as a kind of Puritan scholar fulfilling his calling to labour is in some respects an indication of his own idealised model of what it is to be an ethical person.

This being so, one can begin to appreciate that the emphases in Westfall's biography and in his criticisms of other 'Newtons' are deeply personal. For Westfall to accept Frank Manuel's (1968) portrait of Newton as recklessly dominated by his mother, for example, would be to admit to a less than ideal Newton and its consequences: 'If Newton could be tied to his mother's apron strings, how much more could I – to my mother's or to someone's else's', Westfall asks rhetorically (Westfall 1985: 184). Westfall now endorses the view that biography springs from often unconscious motivations. Before his

interview he set out with the idea that introspection might be an unprofit-
able enterprise but he now sees that any biography is a kind of self-display
(Baron and Pletsch 1985: 10, 13, 16).

The lesson here is surely not to mask interests, nor indeed to let them
overwhelm biography, but to engage in a kind of *disciplined subjectivity*, to
recognise idealisation, ego distortion and transference and see biographical
work as a kind of collaboration between subject and writer. Viewed occasionally as an example of transference (see Edel 1981: 288), this collaboration
has other interesting features that affect the style, aim and accomplishment
of biography. In his chapter, James Moore reflects 'metabiographically' on
the collaboration with his subject, in writing *Darwin* (1991), and also on
that between himself and his co-author Adrian Desmond.

A remarkable contract develops between biographers – and readers – and
the dead figures of the past. There *they* lie – Shakespeare, Newton, Darwin
or whoever – here *we* are desiring to get close to them, and biographers seek
not only to bring them to life but to effect a rapprochement and identification
with them. Moore approves a critic's comment that the skilled biographer
develops a 'split vision', the ability to judge a life with one eye while seeing
with the other how the world appears to his or her subject. In the terms of
Richard Holmes' *Footsteps* (1987), Moore's own quest for Darwin began
with 'identification', the discovery of events and patterns in common with
his subject. Exploring the nature of the common experiences led Moore
further into Darwin's own mind, so far as he was able, in what he calls a
'psychological trespass' (p. 272). The term connotes an illicit intrusion and
forbidden pleasure, feelings seldom admitted in the pages of scholarship. Jacques Roger, historian of science and biographer of Buffon, wrote recently
that it would be interesting 'to submit to deep psychological analysis the
attitude of historians of science with regard to the scientific thinkers that
they study'. Such analysis 'would perhaps reveal some very ambiguous feelings' (Sloan 1994: 469). In Moore's case, the ambiguous feelings produced
a 'haunting' and several 'complications', bound up with the complexities of
Darwin's period and personality. Given that the relationship between biographer and subject continuously reworks both, there are evident difficulties
in maintaining 'historical distance', a clear 'authorial voice' and 'necessary
objectivity'.

How the biographer appears in a biography is a matter of style as much as
of objectivity. On occasion writers intrude in order to remind the reader that
a life story is a construction. For example, in his enormous biography of

Dickens, Peter Ackroyd upbraids his subject for this and that, in the process becoming exasperated, and then pauses to ask himself in an oddly distanced voice '. . . why did you decide to write the book in the first place?' (1990: 895). Into what is in many respects an old-fashioned and straightforward biography, in density of detail and elegant sweep reminiscent of Victorian monuments, Ackroyd insistently intrudes. On one occasion, he conducts an Imaginary Conversation with the reader about his own biography; in another Wilde, Chatterton, Dickens and T.S. Eliot (subjects of other Ackroyd biographies) talk noisily about their lives and biographies; while in a third Ackroyd relates a dream about his subject. Such 'interludes' may be self-indulgent, but have the effect of making the reader think about the possibilities and consequences of immersion in another person's life. Stylistically, what is sparkling and innovative at the outset begins to pale after a while; the narrative is too often interrupted, not because we need to learn by the interruptions about Dickens but because Ackroyd wishes us to learn about himself.

The intrusion of the biographer may be seen as a response to the 'truth-question' in biography. The quest for biographical truth has been thought a hallmark of modern biography; to Woolf — with Strachey in mind — the biographer was 'a votary of Truth, Candour and Honesty . . . the austere Gods who keep watch and ward by the inkpot of biography' (Edel 1978: 142). And questions of truthfulness, accuracy and honesty have been to the fore in biographical criticism. To an extent, such questions can be resolved in technical terms, relating to access to subjects or their papers, a willingness to conceal nothing, to cite sources in full, and so on. There are stories of Boswell torturing himself and his friends to gather a harvest of reminiscences, running across the city to verify this date or that, or retrieve a scrap of paper, all in the search for precious authenticity (Butler 1967: 4). And Richard Evans, in a trenchantly critical review of a recent biography of Himmler, cautioned that 'it is only by tracking down the last scraps of paper, the last piece of evidence, that a rounded picture can be painted' (1990: 899). But this would surely stop biography in its tracks. How would we ever know when to stop research and when to start writing?

However thorough the research, the major effort is that of organisation. 'How', asked Woolf when she sat down to write the life of Roger Fry, 'how can one make a life out of six cardboard boxes full of tailors' bills, lovers' letters and old picture postcards?' (Edel 1981: 2). While it can be suggested that any effort to *organise* a life into a narrative is doomed from the start to the lie that life *is* organised like a work of art, the biographer still needs to decide whether or not to appear in the biography, and in what way, as well as how to offer insight, psychological or artistic. Curiously enough, as biographies have become more subject-revealing, biograph*ers* have become more and more subject-concealing.

Conclusion

Biography touches closely on some of our most central assumptions about the nature of science. On the one hand, it promises an insight into the minds of those few individuals who have made great discoveries about the physical world. It thus engages with long-established religious and philosophical traditions about the character of such persons, their moral integrity, their detachment in intellectual solitude from social and political pressures. On the other hand, it stands in an awkward relation with another powerful ethos of the Western scientific enterprise since the seventeenth century – the social, cooperative nature of scientific thought and method in which the contributions of individuals are tested by others and then, only tentatively, added to the store of knowledge. The tension between these two traditions is visible in two of the anecdotes associated with Isaac Newton, an individual usually claimed unequivocally by the first tradition. 'Does he eat and drink and sleep, is he like other men?', asked a French mathematician, the Marquis de l'Hôpital, thus providing one of many in the set of adulations to Newton's celestial genius. Yet Newton himself, in reportedly describing his achievements as dependent on the 'shoulders of giants' before him, gave some support to the notion of science as a long, collective endeavour (see Fauvel *et al.* 1988: 38, 47; Merton 1965).

This volume of essays does not pretend to resolve this tension, nor even to suggest that it should be attenuated. We hope that the discussion of styles and approaches in literary biography, and the reflections of some contributors who have written scientific biographies, will provide further material for debate on this topic. Alternatively, for historians who do not intend to embark on a biography, this introduction and some of the chapters may stimulate further consideration of what Dorinda Outram invited some time ago: a heightening of 'our awareness of the role which biographical inquiry has played in the history of science' (Outram 1976: 101). The most common understanding of this concerns the way conceptions of genius, deployed in biographies, have competed with explanations in terms of abstract, social and historical factors ranging from material conditions to political ideology. But some recent scholarship suggests that the gulf need not be so wide.

In the period from the Scientific Revolution to the late nineteenth century there was an explicit concern with the character, motivations and values of the persons engaged in the pursuit of natural knowledge. The social legitimation of the nascent scientific enterprise, and its bid for increased cultural status in the nineteenth century, depended heavily on claims about the kind of people it attracted. Steven Shapin's recent work shows how a biographical focus on Robert Boyle can move to an historical analysis of social norms: Boyle's life can be appreciated as an exemplar of what it meant to be a natural philosopher in the early modern period (Shapin 1994: especially chapter 4, also 1993; for another perspective, see Hunter 1995).

Interestingly, this approach seems to qualify the contrast drawn by Merton between the moral integrity of individual scientists (which he regarded as outside sociological inquiry) and the conventions and methods of the scientific community. On Shapin's account, individual models of a moral life were crucial to the historical formation of the scientific ethos – partly expressed in Merton's 'normative structure of science'. Even in the nineteenth century it is arguable that intellectual authority was still embodied in individuals, although this gave way as knowledge was increasingly presented as certified by shared method and institutions (see Outram 1984; Yeo 1993: chapter 5). Given this, there is a need to recover issues that might have been previously dismissed as relating to the 'merely personal'. Thus other work on the way in which individual men and women have made vocations for themselves in science in various historical periods also opens a role for biography which historians of science might embrace (Sonntag 1974; Abir-Am and Outram 1987; Shapin 1989; Schweber 1989; Morus 1992; Oster 1993). These suggestions do not simply point to a new defence of biography, but to the importance of personal identities and life histories in the cultural history of science.

Bibliography

Aaron, D. (ed.) (1978) *Studies in Biography*. Cambridge, MA: Harvard University Press.
Abir-Am, P. (1991) Noblesse oblige: biographical writings on Nobelists. *Isis*, **82**, 326–43.
Abir-Am, P. and Outram, D. (eds.) (1987) *Uneasy Careers and Intimate Lives: Women in Science, 1789–1979*. New Brunswick: Rutgers University Press.
Ackroyd, P. (1990) *Dickens*. London: Sinclair-Stevenson.
Anon. (1879*a*) Contemporary literature: biography, travel and sport. *Blackwood's Edinburgh Magazine*, **125**, 482–506.

Anon. (1879*b*) Studies in biography. *Fraser's Magazine,* 596, 255–75.

Bacon, F. (1974) *The Advancement of Learning and New Atlantis* (ed. A. Johnston). Oxford: Clarendon Press.

Bagehot, W. (1965–86) *The Collected Works of Walter Bagehot* (ed. N. St John-Stevas), 15 vols. London: *The Economist.*

Banville, J. (1976) *Doctor Copernicus.* London: Secker and Warburg.

Banville, J. (1981) *Kepler.* London: Secker and Warburg.

Baron, S.H. and Pletsch, C. (eds) (1985) *Introspection in Biography: The Biographer's Quest for Self-Awareness.* Hillside, NJ: Analytic Press.

Barthes, R. (1977) The death of the author. In R. Barthes, *Image, Music, Text.* New York: Hill and Wang, pp. 142–8.

Bayne, P. (1871) *The Life and Letters of Hugh Miller,* 2 vols. London: Strahan and Co.

Bazerman, C. (1991) How natural philosophers can cooperate: the literary technology of coordinated investigation in Joseph Priestley's *History and Present State of Electricity.* In C. Bazerman and J. Paradis (eds), *Textual Dynamics of the Professions.* Madison: University of Wisconsin Press, pp. 13–44.

Beauchamp, E. (1990) Education and biography in the contemporary period. *Biography,* 13, 1–5.

Bernard, C. (1957) [1865] *An Introduction to the Study of Experimental Medicine,* trans. H.C. Greene. New York: Dover.

Bowker, G. (1993) The age of biography is upon us. *Times Higher Education Supplement,* 8 January, 19.

Bowlby, J. (1990) *Charles Darwin: A New Life.* London: Hutchinson.

Brewster, D. (1831) *The Life of Isaac Newton.* London: J. Murray.

Brewster, D. (1855) *Memoirs of the Life, Writings and Discoveries of Sir Isaac Newton,* 2 vols. Edinburgh: Constable.

Broderick, D. (1994) *The Architecture of Babel: Discourses of Literature and Science.* Melbourne: Melbourne University Press.

Browne, J. (1978) The Charles Darwin–Joseph Hooker correspondence: an analysis of manuscript resources and their use in biography. *Journal of the Society for the Bibliography of National History,* 18, 351–66.

Browne, J.E. (1981) The making of the *Memoir* of Edward Forbes. *Archives for Natural History,* 10, 205–19.

Burke, P. (1991) Reflections on the origin of cultural history. In J.H. Pittock and A. Wear (eds.), *Interpretation and Cultural History.* New York: St Martin's Press, pp. 153–74.

Butler, R.A. (1967) *The Difficult Art of Autobiography.* Oxford: Clarendon Press.

Candolle, A. de (1873) *Histoire des sciences et des savants depuis deux siècles.* Geneva: H. Georg.

Cantor, G.N. (1991) *Michael Faraday: Sandemanian and Scientist. A Study of Science and Religion in the Nineteenth Century.* London: Macmillan.

[Carlyle, T.] (1832) Biography. *Fraser's Magazine,* 5, 253–60.

[Carlyle, T.] (1838) Memoirs of the life of Scott. *The Westminster Review,* 6, 293–345.

[Carlyle, T.] (1885) *Critical and Miscellaneous Essays.* New York: John and Alden.

Carr, H.W. (1929) *Leibniz.* Boston: Little, Brown.

Charmley, J. and Homberger, E. (eds) (1988) *The Troubled Face of Biography.* London: Macmillan.

[Christie, R.H.] (1884) biographical dictionaries. *Quarterly Review*, 157, 187–230.

Clifford, J.L. (1962) *Biography as an Art*. Oxford: Oxford University Press.

Clifford, J.L. (1970) *From Puzzles to Portraits*. Chapel Hill, NC: University of North Carolina Press.

Clifford, J.L. (1978) 'Hanging up looking glasses at odd corners': ethnobiographical prospects. In Aaron (1978), pp. 41–56.

Cockshut, A.O.J. (1984) *The Art of Autobiography in Nineteenth and Twentieth Century England*. New Haven: Yale University Press.

Cockshut, A.O.J. (1987) Foreword. In Hoberman, pp. ix–xiv.

Colp, R. (1977) *To Be an Invalid: the Illness of Charles Darwin*. Chicago: University of Chicago Press.

Colp, R. (1987) Charles Darwin's insufferable grief. *Free Associations*, 9, 7–44.

Cox, J.M. (1987) Review of David Novarr, *The Lines of Life*. *Biography*, 10, 370–2.

Crook, N. and Guiton, D. (1986) *Shelley's Venomed Melody*. Cambridge: Cambridge University Press.

D'Alembert, J. (1963) *Preliminary Discourse to the Encyclopedia of Diderot*, trans. R. Schwab. Indianapolis: Bobbs-Merrill.

Darwin, C. (1974) *Charles Darwin and T.H. Huxley Autobiographies* (ed. G. de Beer). London: Oxford University Press.

Daston, L. (1992) Objectivity and the escape from perspective. *Social Studies of Science*, 22, 597–618.

Delany, P. (1969) *British Autobiography in the Seventeenth Century*. London: Routledge.

De Morgan, A. (1855) Sir David Brewster's *Life of Newton*. *North British Review*, 23, 307–38.

Desmond, A. and Moore, J. (1991) *Darwin*. London: Michael Joseph.

Dobbs, B.J.T. (1994) Review of A. Rupert Hall, *Isaac Newton* and Richard Westfall, *Life of Isaac Newton*. *Isis*, 85, 515–16.

D.S. (1994) NB. *The Times Literary Supplement*, 4 November, 16.

Edel, L. (1978) *Literary Biography*. Bloomington: Indiana University Press.

Edel, L. (1981) Biography and the science of man. In Friedson 1981, pp. 1–11.

Ehrenpreis, I. (1962–83) *Swift: the Man, His Work and the Age*, 3 vols. Cambridge, MA: Harvard University Press,

Einstein, A. (1949) Autobiographical notes. In P.A. Schipp (ed.), *Albert Einstein: Philosopher Scientist*. Evanston, IL: Library of Living Philosophers, pp. 3–95.

Elliott, C.A. (1982) Models of the American scientist: a look at collective biography. *Isis*, 73, 77–93.

Elliott, C.A. (1990) Collective lives of American scientists: an introductory essay and a bibliography. In E. Garber (ed.), *Beyond History of Science: Essays in Honor of Robert E. Schofield*, Bethlehem, PA: Lehigh University Press, pp. 81–104.

Enright, D.J. (ed.) (1989) *Ill at Ease: Writers on Ailments Real and Imagined*. London: Faber and Faber.

Epstein, W.H. (1987) *Recognizing Biography*. Philadelphia: University of Pennsylvannia Press.

Fauvel, J., Flood, R., Shortland, M. and Wilson, R. (eds) (1988) *Let Newton Be! A New Perspective on His Life and Work*. Oxford: Clarendon Press.

Flanagan, T. (1982) Problems of psychobiography. *Queen's Quarterly*, 89, 596–610.

Freud, S. (1959) *The Standard Edition of the Complete Psychological Works of Sigmund Freud*, vol. 20. London: Hogarth Press, 7–74.

Friedson, A.M. (ed.) (1981) *New Directions in Biography*. Manoa: University Press of Hawaii.

Fuller, S. (1944) Can science studies be spoken in a civil tongue? *Social Studies of Science*, 24, 143–68.

Gagnier, R. (1991) *Subjectivities: A History of Self-representation in Britain 1832–1920*. Oxford: Oxford University Press.

Galton, F. (1869) *Hereditary Genius: an Enquiry into its Laws and Consequences*. London: Macmillan.

Garfield, E. (1990) Scientific biography. *Current Contents*, 29 January, 3–5.

Garraty, J.A. (1957) *The Nature of Biography*. New York: Alfred Knopf.

Gaukroger, S. (1995) *Descartes, An Intellectual Biography*. Oxford: Oxford University Press.

Geison, G. (1978) *Michael Foster and the Cambridge School of Physiology: the Scientific Enterprise in Late Victorian Society*. Princeton: Princeton University Press.

Gittings, R. (1978) *The Nature of Biography*. Seattle: Washington University Press.

Goldstein, J.H. (1989) Darwin, Chagas, mind and body. *Perspectives in Biology and Medicine*, 32, 586–600

Gross, A.R. (1990) *The Rhetoric of Science*. Cambridge, MA: Harvard University Press.

Gross, A.R. (1991) Rhetoric of science without constraints. *Rhetorica*, 9, 283–316.

Gross, P.R. and Levitt, N. (1994) *Higher Superstition: The Academic Left and its Quarrel with Science*. Baltimore: The Johns Hopkins University Press.

Hamilton, I. (1994) *Walking Possession: Essays and Reviews, 1968–93*. London: Bloomsbury.

Hankins, T.L. (1979) In defence of biography: the use of biography in the history of science. *History of Science*, 17, 1–16.

Harris, J. (1993) *Private Lives, Public Spirit: A Social History of Britain 1870–1914*. Oxford: Oxford University Press.

Harrison, F. (ed.) (1892) *The New Calendar of Great Men: Biographies of the 558 Worthies of all Ages and Nations in the Positivist Calendar of Auguste Comte*. London: Macmillan.

Hoberman, R. (1987) *Modernizing Lives: Experiments in English Biography, 1918–1939*. Carbondale: Southern Illinois University Press.

Hoffman, L.E. (1984) Early psychobiography, 1900–1930: some reconsiderations. *Biography*, 7, 341–51.

Holmes, L.F. (1974) *Claude Bernard and Animal Chemistry: the Emergence of a Scientist*. Cambridge, MA: Harvard University Press.

Holmes, L.F. (1981) The fine structure of scientific creativity. *History of Science*, 19, 60–70.

Holmes, R. (1990) Time's golden handshake. *Guardian Weekly*, 30 December, p. 21.

Holroyd, M. (1974) *Unreceived Opinions*. New York: Holt, Rinehart and Winston.

Holroyd, M. (1981) Literary and historical biography. In Friedson, pp. 12–25.

Horney, K. (1937) *Neurotic Personality of Our Time*. New York: Morrow.

Howard, M.W. (1970) *The Influence of Plutarch in the Major European Literature of the Eighteenth Century*. Chapel Hill, NC: University of North Carolina Press.

Hume, D. (1993) *Selected Essays* (eds S. Copley and A. Edgar). Oxford: Oxford University Press.

Hunter, M. (1995) How Boyle became a scientist. *History of Science*, 33, 59–103.

Huxley, T.H. (1974) *Charles Darwin and T.H. Huxley Autobiographies* (ed. G. de Beer). London: Oxford University Press.

Ireland, N.O. (1962) *Index to Scientists of the World from Ancient to Modern Times: Biographies and Portraits*. Boston: F.W. Faxton.

Jacob, F. (1988) *The Statue Within: An Autobiography*. New York: Basic Books.

Kaplan, J. (1978) The 'real life'. In Aaron, (1978), pp. 1–8.

Kaplan, J. (1994) A culture of biography. *The Yale Review*, **82**, 1–16.

Kendall, P.M. (1965) *The Art of Biography*. London: George Allen and Unwin.

Kingsbury, M. (1987) Congenial associates: the biographical essays of William Osler. *Biography*, **10**, 225–40.

Knight, D.M. (1992) *Humphry Davy: Science and Power*. Oxford: Blackwell.

Kragh, H. (1987) *An Introduction to the Historiography of Science*. Cambridge: Cambridge University Press.

Laertius, D. (1954) *Lives of Eminent Philosophers*, with a trans. by R.D. Hicks. Cambridge, MA: Heinemann.

Laudan, R. (1993) Histories of the sciences and their uses: a review to 1913. *History of Science*, **31**, 1–34.

[Lewes, G.H.] (1878) On the dread and dislike of science. *Fortnightly Review*, **29**, 805–15.

Lewontin, D.C. (1985) Darwin, Mendel and the mind. *New York Review of Books*, 10 October, 18–23.

Locke, D. (1992) *Science as Writing*. New Haven: Yale University Press.

Lyell, C. to G. Mantell. Unpublished letter of 23 April 1830. In Gideon Mantell Collection, Alexander Turnbull Library, Wellington, New Zealand, folder 62.

McEvoy J. (1979) Electricity, knowledge and the nature of progress in Priestley's thought. *British Journal for the History of Science*, **5**, 1–30.

Mandell, G.P. (1991) *Life into Art: Conversations with Seven Contemporary Biographers*. Fayetteville: The University of Arkansas Press.

Manuel, F. (1968) *A Portrait of Isaac Newton*. Cambridge: Harvard University Press.

Markus, G. (1987) Why is there no hermeneutics of natural sciences? Some preliminary theses. *Science in Context*, **1**, 5–51.

Medawar, P. (1964) Is the scientific report fraudulent? Yes; it misrepresents scientific thought. *Saturday Review*, **47**, 1 August, 42–43.

Merton, R.K. (1938) Science, technology and society in seventeenth-century England. *Osiris*, **4**, part 2, 360–632.

Merton, R.K. (1965) *On the Shoulders of Giants: a Shandean Postscript*. New York: The Free Press.

Merton, R.K. (1973) *The Sociology of Science: Theoretical and Practical Investigations* (ed. N.W. Storer). Chicago: University of Chicago Press.

Mikulinsky, S. (1974) Alphonse de Condolle's *Histoire des sciences et des savants depuis deux siècles* and its historical significance. *Organon*, **10**, 223–43.

Moore, W. (1989) *Schrödinger: Life and Thought*. Cambridge: Cambridge University Press.

Morton, L.T. and Moore, R.J. (1989) *A Bibliography of Medical and Biomedical Biography*. Aldershot: Scolar Press.

Morus, I.R. (1992) Different experimental lives: Michael Faraday and William Sturgeon. *History of Science*, **30**, 1–28.

Nadel, I.B. (1984) *Biography: Fiction, Fact and Form*. London: Macmillan.

Oliphant, M.O.W. (1883) The ethics of biography. *Contemporary Review*, 44, 76–93.

Oster, M. (1993) Biography: culture, and science: the formative years of Robert Boyle. *History of Science*, 31, 177–226.

Outram, D. (1976) Scientific biography and the case of Georges Cuvier: with a critical bibliography. *History of Science*, 14, 101–37.

Outram, D. (1978) The language of natural power: the *éloges* of Georges Cuvier. *History of Science*, 16, 153–78.

Outram, D. (1984) *Georges Cuvier. Vocation, Science and Authority in Post-revolutionary France*. Manchester: Manchester University Press.

Outram, D. (1989) *The Body and the French Revolution: Sex, Class and Political Culture*. New Haven: Yale University Press.

Owen, R. (1894) *The Life of Richard Owen, by his Grandson*, 2 vols. New York: D. Appleton.

Pachter, M. (ed.) (1979) *Telling Lives: The Biographer's Art*. Washington, D.C.: New Republic.

Paul, C.B. (1980) *Science and Immortality: the Eloges of the Paris Academy of Sciences (1699–1791)*. Berkeley: University of California Press.

Playfair, J. (1815–24) Dissertation second: exhibiting a general view of the progress of mathematical and physical science, since the revival of letters in Europe. *Supplement to the Encyclopaedia Britannica* (ed. M. Napier), 6 vols. Edinburgh: Constable, vol. 2 and vol. 4.

Porter, R. (1996) Miller's madness. In M. Shortland (ed.) *Hugh Miller and the Controversies of Victorian Science*. Oxford: Oxford University Press, pp. 285–309.

Priestley, J. (1765) *A Description of a Chart of Biography, with a Catalogue of Names Inserted in it, and the Dates Annexed to Them*. Warrington: the Author.

Priestley, J. (1775) *The History and Present State of Electricity, with Original Experiments*, 3rd. edition, 2 vols. London: C. Bathurst and T. Lowndes.

Purcell, E.S. (1896) On the ethics of suppression in biography. *Nineteenth Century*, 40, 533–42

Pyenson, L. (1977) 'Who the guys were': prosopography in the history of science. *History of Science*, 15, 155–88.

Rodden, J. (1989) *The Politics of Literary Reputation: the Making and Claiming of 'St. George' Orwell*. New York: Oxford University Press.

Schaffer, S. (1986) Scientific discoveries and the end of natural philosophy. *Social Studies of Science*, 16, 387–420.

Schweber, S. (1989) John Herschel and Charles Darwin: a history in parallel lives. *Journal of the History of Biology*, 22, 1–72.

Shapin, S. (1989) Who was Robert Hooke? In M. Hunter and S. Schaffer (eds), *Robert Hooke: New Studies*. Woodbridge, Suffolk: Boydell Press, pp. 253–85.

Shapin, S. (1991) A scholar and a gentleman: the problematic identity of the scientific practitioner in early modern England. *History of Science*, 29, 279–327.

Shapin, S. (1993) Personal development and intellectual biography: the case of Robert Boyle. *British Journal for the History of Science*, 26, 335–45.

Shapin, S. (1994) *A Social History of Truth: Civility and Science in Seventeenth-Century England*. Chicago: University of Chicago Press.

Shapin, S. and A. Thackray. (1974) Prosopography as a research tool in history of science: the British scientific community 1700–1900. *History of Science*, 12, 1–28.

Sheets-Pyenson, S. (1990) New directions for scientific biography: the case of Sir William Dawson. *History of Science*, **28**, 399–410.

Sheringham, M. (1994) *French Autobiography: Devices and Desires*. Oxford: Clarendon Press.

Shore, M.F. (1980) Biography in the 1980s: a psychoanalytic perspective. *Journal of Interdisciplinary History*, **12**, 89–113.

Shortland, M. (1988) Exemplary lives: a study of scientific autobiographies. *Science and Public Policy*, **15**, 170–9.

Shortland, M. (1995) Powers of recall: Sigmund Freud's partiality for the prehistoric. *Australian Journal of Historical Archaeology*, **11**, 3–20.

Shortland, M. (1996) Bonneted mechanic and narrative hero: the self-modelling of Hugh Miller. In M. Shortland (ed.) *Hugh Miller and the Controversies of Victorian Science*. Oxford: Oxford University Press, pp. 14–86.

Skidelsky, R. (1987) Exemplary lives. *The Times Literary Supplement*, November, 1250.

Sloan, P. (1994) Buffon studies today. *History of Science*, **32**, 469–77.

Smith, C. and Wise, N. (1989) *Energy and Empire: a Biographical Study of Lord Kelvin*. Cambridge: Cambridge University Press.

Smith, F. (1990) Charles Darwin's ill health. *Journal of the History of Biology*, **23**, 443–59.

Smith, F. (1992) Charles Darwin's health problems: the allergy hypothesis. *Journal of the History of Biology*, **25**, 285–306.

Sonntag, O. (1974) The motivations of the scientist: the self-image of Albrecht von Haller. *Isis*, **64**, 336–51.

Sokal, M. (1990) Life-span developmental psychology and the history of science. In E. Garber (ed.), *Beyond History of Science: Essays in Honor of Robert E. Schofield*, Bethlehem, PA: Lehigh University Press, pp. 67–80.

Sokoloff, B. (1938) *Napoleon: a Medical Approach*. London: Selwyn & Blount.

Stanfield, J.F. (1813) *An Essay on the Study and Composition of Biography*. Sunderland: George Garbutt.

Stanley, T. (1701) *The History of Philosophy; Containing the Lives, Opinions, Actions, and Discourses of the Philosophers of Every Sect*, 3rd edition. London: W. Battesby.

Stauffer, D.A. (1930) *English Biography Before 1700*. Cambridge: Cambridge University Press.

Stauffer, D.A. (1941) *The Art of Biography in Eighteenth Century England*. New Jersey: Princeton University Press.

Stein, D. (1985) *Ada: a Life and a Legacy*. Cambridge, MA: MIT Press.

Stephen, L. (1893) Biography. *National Review*, **22**, 171–83.

Strouse, J. (1978) Semiprivate lives. In Aaron, pp. 113–29.

Strouse, J. (1983) Response. In Veninga, pp. 37–41.

Sturges, P. (1983) Collective Biography in the 1980s. *Biography*, **6**, 316–32.

Sulloway, F.J. (in preparation) *Born to Rebel: Radical Thinking in Science and Social Thought*.

Todhunter, I. (1876) *William Whewell, D.D.*, 2 vols. London: Macmillan.

Theerman, P. (1985) Unaccustomed role: the scientist as historical biographer – two nineteenth-century portrayals of Newton. *Biography*, **8**, 145–62.

Theerman, P. (1990) National images of science: British and American views of scientific heroes in the early nineteenth century. In J.W. Slade and J.Y. Yee (eds), *Beyond the Two Cultures: Essays on Science, Technology and Literature*, Iowa: Iowa State University Press.

Veninga, J.F. (ed.) (1983) *The Biographer's Gift: Life Histories and Humanism*. College Station: Texas A&M Press.

Vidal, F. (1994) *Piaget before Piaget*. Cambridge, MA.: Harvard University Press.

Ward, W. (1896) Candour in Biography. *New Review*, 14, 445–52.

Weintraub, S. (1987) *Victoria: An Intimate Portrait*. New York: Dutton.

Weisz, G. (1988) The self-made mandarin: the *Eloges* of the French Academy of Medicine, 1824–47. *History of Science*, 26, 13–40.

Westfall, R.S. (1980) *Never at Rest: A biography of Isaac Newton*. Cambridge: Cambridge University Press.

Westfall, R.S. (1985) Newton and his biographer. In Baron and Pletsch, pp. 175–89.

Whewell, W. (1857 [1837]) *History of the Inductive Sciences*, 3rd edition, 3 vols. London: J. Parker.

Williams, L.P. (1991) The life of science and scientific lives. *Physis*, 28, 199–213.

Williams, T. (ed.) (1994) *The Collins Biographical Dictionary of Scientists*. London: Collins.

Wilson, L. (1977) The puzzling illness of Charles Darwin. *Journal of the History of Medicine and Allied Sciences*, 32, 437–42.

Wiltshire, J. (1991) *Samuel Johnson in the Medical World: the Doctor and the Patient*. Cambridge: Cambridge University Press.

Wolpert, L. (1992) *The Unnatural Nature of Science: Why Science Does Not Make (Common) Sense*. London: Faber.

Yeo, R. (1988) Genius, method and morality: images of Newton in Britain, 1760–1860. *Science in Context*, 2, 257–84.

Yeo, R. (1993) *Defining Science: William Whewell, Natural Knowledge and Public Debate in Early Victorian Britain*. Cambridge: Cambridge University Press.

Young, R.M. (1987) Darwin and the genre of biography. In G. Levine (ed.), *One Culture: Essays in Science and Literature*. Madison: University of Wisconsin Press, 203–24.

Young, R.M. (1988) Biography: the basic discipline for human sciences. *Free Associations*, 9, 108–30.

Young-Bruehl, E. (1993) Biographies of Freud. In M. Micale and R. Porter (eds.), *Discovering the History of Psychiatry*. Oxford: Oxford University Press.

Zoubov, V. (1962) L'histoire de la science et la biographie des savants. *Kwartalnik Historii Nauki i Techniki*, 6, 29–42.

1

Existential projects and existential choice in science: science biography as an edifying genre

THOMAS SÖDERQVIST

All the people of this lonely world, have a piece of pain inside.
(Eurhythmics, 'When the day goes down')

Introduction

During the last decade an increasing number of high quality biographies of scientists have appeared on the book market[1] — Richard Westfall's Newton study, *Never at Rest*, William Provine's *Sewall Wright and Evolutionary Biology*, Crosbie Smith and Norton Wise's study of Lord Kelvin and Victorian England, David Cassidy's Heisenberg biography, Geoffrey Cantor's study of Faraday, Adrian Desmond and James Moore's Darwin tome, and Frederic Holmes's first volume on Hans Krebs — just to name some of the most admirable works.[2] Athough still within the traditional confines of the genre, these and similar biographies are more detailed, better researched, more stylishly written, and more penetrating than almost any biography written just a generation ago. Each new biography seems to be unrivalled. For someone who browses through the history of science shelves of an academic bookstore these works indicate that science biography stands out as a most — if not *the* most — impressive genre of the discipline.

---

[1] I use the expressions 'science biography' and 'biographies of scientists' instead of 'scientific biography', partly because 'scientific biography' implies a bias in favour of the scientific activities as opposed to other activities in life, and partly because it has a built-in ambiguity ('scientific' as opposed to 'unscientific').

[2] Westfall (1980), Provine (1986), Smith and Wise (1989), Cassidy (1991), Cantor (1991), Desmond and Moore (1991), Holmes (1991).

In spite of the recent flourishing state of science biography, however, there is a widespread ambivalence and uncertainty as to the role and place of biography among historians of science. Biographical studies have dominated the history of science for most of its existence: whether cast in the form of life-and-times monographs, or in the form of studies of a scientist's contribution to the history of an idea or to the creation of an institution, biography was a universally respected and unproblematic genre. The eighteen volume *Dictionary of Scientific Biography* stands out as a testimony of this classical age of science biography. But while its popular attraction remains unshattered, its traditional privileged status and appeal in academia have been under siege during the last decades. Emmanuel Le Roy Ladurie referred as much to science biography when stating in the late 1970s that '[p]resent-day historiography, with its preference for the quantifiable, the statistical and the structural, has . . . virtually condemned to death . . . the individual biography'.[3] Its loss of academic status has repeatedly been regretted throughout the past decade, from Thomas Hankins, who noticed that 'modern trends in the history of science seem to leave little room for biography',[4] over Helge Kragh's reference to the 'diminishing respectability of the biography',[5] to Michael Sokal's recent conclusion that there exists a widespread scepticism about the value of biographical inquiry.[6]

Hence, anyone who sets out to write a biography of a scientist these days can hardly avoid being confronted with a number of questions concerning the aims of the genre. What is the legitimate place of biography in history of science? Is it simply a sort of sophisticated entertainment, the scientist's bedside companion after the daily torments in the laboratory or at the desk, and thus better handed over to novelists, or is biography a possible and valuable scholarly pursuit in itself? If so, is it primarily an aid for the history of science, a tool for understanding the succession of theories or ideas of a certain time, or a looking glass through which we can investigate institutional structures or the social construction of scientific knowledge? Is it a generator of cases for the philosophy, psychology or sociology of science that may help us explain the origin of theories or the problem of creativity? Or will biographical narratives be able to fulfil more fundamental needs, even providing

---

[3] Le Roy Ladurie (1979), 111.
[4] Hankins (1979), 3.
[5] Kragh (1989), 168.
[6] Sokal (1990).

exemplars through which we can learn to tackle the existential problems we confront in our intellectual lives?

In this chapter I discuss these and sim. lar recent challenges to the genre, with the intention of formulating an argument for an existential approach to biographies of scientists. My argument, based on experience from my research during the last couple of years for a biography of a leading contemporary immunologist, Niels K. Jerne (1911–1994), is that the aim of biography is not primarily to be an aid for the history of science, nor to be a generator of case studies. Instead of adding to the 'hermeneutics of suspicion' that governs so much of today's history and sociology of science, the main purpose of science biography is, I suggest, as a genre that can provide a variety of exemplars of existential projects of individual scientists – narratives through which we can identify ourselves with others who have been confronted with existential choices and struggled with the existential conditions for living in and with science. Such life stories not only provide us with opportunities to understand ourselves, intellectually as well as emotionally, but may also change and create ourselves. Hence, biographies of scientists are 'edifying' – they can help us reorient our familiar ways of thinking about our lives in unfamiliar terms, and 'take us out of our old selves by the power of strangeness, to aid us in becoming new beings'.[7]

The sociological redefinition of science biography: social biography as an auxiliary to the social history of science

The prevaling uncertainty about science biography is to a large extent the result of an increasing uneasiness over the years among historians of science about dealing with the personality of the individual scientist. Thus, three successive waves of suspicion against (even dismissal of) biography and the scientist as a person can be identified: one philosophical, the other sociological, and the third post-structuralist. The first, indirect, blow against the genre was a result of the merger between history and philosophy of science by which the historiography of science became increasingly influenced by philosophers who emphasised the logical structure of scientific ideas and disregarded the importance of relating it to the scientist. The individual scientist was not ignored as such – for example, I. B. Cohen considered Alexandre

---

[7] Rorty (1980), 360.

Koyré's *Etudes galiléennes* to be a brilliant attempt to go behind formal presentation and 'to understand the mind and thinking process of an important scientist',[8] and the individual scientist was frequently used as case-material for the rational reconstruction of the history of ideas and research programmes. However, all but pure cognition – the personality, the passions, and the idiosyncratic aspects of scientific work – was squeezed out by this joint history and philosophy of science programme. The life, particularly the personal, embodied, life of the scientist, was taken to be irrelevant for the understanding of science, as if public faces in private places were nicer and wiser than private faces in public places. This was an attitude reinforced by the privileged role attached to the 'context of justification' as compared with studies of the 'context of discovery'.[9] Distracting voices, such as Michael Polanyi's, which reminded fellow philosophers about the passionate nature of 'private' and 'tacit' knowledge, did not have any impact on the history of science or science biography.[10]

This view was, and is probably still, supported by many scientists themselves. For example, Albert Einstein, at least publicly, showed a disdain of the 'merely personal' aspects of his life,[11] quite similar to Hannah Arendt's opinion that biography 'is rather unsuitable for . . . the lives of artists, writers, and generally men and women whose genius forced them to keep the world at a certain distance and whose significance lies chiefly in their works, the artifacts they added to the world, not the role they played in it'.[12] It should be noted that this rejection of the personal and biographical aspects of the intellectual life has an interesting parallel in literary history and criticism. Authors and literary critics, such as Paul Valéry and Marcel Proust in France and T.S. Eliot in England laid the groundwork for the anti-biographical programme of New Criticism, whose proponents rejected the earlier strong programme of biographical writing and claimed that an understanding of the state of mind of the author was of no use for understanding a work of art. What interests us, said Eliot, is the inner composition of the work of art, its style, symbolism, and so forth. The 'objective correlative', not the author behind it, can move the reader into a particular state of

---

[8] Cohen (1987), 55–6.
[9] In his autobiography, Popper (1974, 47) describes a traumatic youth experience which 'ultimately led even to my distinction between world 2 and world 3'. He does not, however, reflexively consider the consequences of this autobiographical understanding for his later philosophy.
[10] Polanyi (1958) and (1966).
[11] Bernstein (1985).
[12] Young-Bruehl (1982), xvi.

mind.[13] Similar repudiations of biography for being too focused on the personal and individual have later been made by art historians.[14]

During the last two decades the philosophically inspired history of science has gradually been replaced by a concern for the social and political context of science. Not to the advantage of biography, however, for the philosophical dismissal of the personal has been followed by a sociological dismissal of the individual. The genre of biography has been challenged by social historians and sociologists who consider studies of individual scientists and their personalities to be largely irrelevant for the history of scientific disciplines, research schools, and scientific societies, or for understanding the social construction of scientific knowledge. Richard Lewontin summarises the social historical point of view when he warns against the danger that 'by concentrating on the individual creators of ideas or fashions, one may easily fail to ask what social circumstances engendered the problematic in the first place'.[15] The merger between history and sociology of knowledge[16] has further strengthened the doubts about the value of biography. Not even Charles Rosenberg's cautious plea for an actor-oriented approach to history of science has been accepted by the more hard-nosed sociologically oriented historians:[17] Steven Shapin warns against '[t]he risk . . . that the admirable historical goal of understanding actors' categories can wind up dissolving the subject-matter of history of science into atomising particularism'. Instead, he continues, 'the individualistic reflexes that characterize much history may be usefully disciplined by the sociologist's collectivism'.[18] Likewise the recent turn towards discourse analysis and rhetorics of science has, by concentrating on the text, further weakened the interest in the individual scientist and severed the work from the author.[19]

[13] Eliot (1960), 100.
[14] See, for example, Krauss's (1985) attack on 'art as autobiography', that is the view that paintings are expressions of the life of the artist.
[15] Lewontin (1986).
[16] Golinski (1990).
[17] Rosenberg (1988).
[18] Shapin (1992), 354–5. Likewise Krauss (1985) finds the richness of art history in all the different 'ways of understanding art in transpersonal terms: ways that involve questions of period, style, of shared formal and iconographic symbols that seem to be the function of larger units of history than the restricted profile of a merely private life' (p. 25), and believes that an art history that turns 'militantly' away from all that is transpersonal in history, i.e., 'style, social and economic context, archive, structure', is symbolised by an art history 'as a history of proper names'. By 'proper names' Krauss means the tendency among art historians to interpret works of art as representing concrete persons, and the 'art as biography' interpretation amounts to showing whom the artist had in mind as a model when painting. Krauss' main objection to this practice of 'positive identification' of the picture with an identifiable person is that it restricts the space of interpretations.
[19] Smocovitis (1991).

The sociological challenge to science biography has been particularly strong from the side of structuralists such as Fritz Ringer, who criticises what he considers to be an implicit methodological individualism in intellectual history and history of science, rejects the search for the 'subjective project' of the intellectual agent, and insists that it is imperative to disregard the authorial intentions of individual texts.[20] He asserts that intellectual fields 'are entities in their own right, that must not be reduced to aggregates of individuals', and, stressing the need to understand a great text 'positionally' by understanding its relationships to an intellectual field, he advocates instead studies of the relation between the text and 'an existing field of other texts':[21] We must learn 'to understand a cluster of texts as a whole, or as a set of relationships, rather than as a sum of individual statements'.[22] While not rejecting the genre of biography altogether, Ringer relegates it to a secondary, and (to my best understanding) impossible role: 'I believe that biographies are more difficult to write than surveys of intellectual fields, and that they are likely to fail, unless they can draw upon prior investigations of their fields'.[23] Ringer's position (or rather, to follow his own recommendation, the quoted position in the network of relationships, i.e. the proper name 'Fritz Ringer') is extreme, but at the same time consistent. His is a clear top-down view of the relation between structural history and biography: biographies makes sense only when you have identified the positions in the intellectual field; therefore, first map the field, then (perhaps) write biographies.

The impact of social historical and sociological approaches to history of science has certainly had its positive effects on the genre of science biography. The biographers of the 1980s are much more aware of the cultural, social, and political context of the lives of their subjects than were biographers of earlier generations, thereby implicitly endorsing Thomas Mann's view that a man lives not only his personal life, as an individual, but also, consciously or unconsciously, the life of his epoch and of his contemporaries. Indeed, the value of biography as a means for demonstrating the social context of science is the most common argument for the use of biography in our day. Through the life story of the individual scientist we are supposed to understand the culture and the time: 'The historical biographer tries to see through the personality to obtain a better understanding of contemporary

---

[20] Ringer (1990); Ringer is primarily inspired by Pierre Bourdieu (1975).
[21] Ringer (1990), 277, 272.
[22] Ibid., 275.
[23] Ibid.

events and ideas', writes Hankins;[24] 'most importantly, biographies can be used for the intellectual history of the times in which they are written', adds Paul Theerman.[25] But there is a negative side-effect of the impact of social historical and sociological approaches in that many seem to believe that to provide cases for the interplay between social, political and other factors is the only use there is of the genre. For example, Robert M. Young advocates biography as the genre *par préférence* for demonstrating the contextualisation and historicity of science,[26] and Pnina Abir-Am, in a recent critical evaluation of science (auto)biographies, finds these works useful only if they 'reflect awareness of the social, political, and cultural context', illustrate 'gender assymetry', help focus on 'intermediary units of sociohistorical analysis' or illuminate phenomena such as 'the rise of new sociocognitive hierarchies'.[27] Hence, the individual is reduced to a mere instance in contextual history.

As a consequence, science biography has become an ambiguous genre with regard to the role of the individual and the personality in historical narrative. This ambiguity has repeatedly been expressed programmatically during the last decade, for example, by Evelyn Fox Keller, who maintains that a biographical portrait is 'always' done 'against the background of the community' and that biographies 'of necessity' must serve 'simultaneously as biography and as intellectual history'.[28] Similarly, while advocating an actor-oriented approach to history of science, including a sensitivity to the individual's choices during his life-course, Rosenberg nevertheless wants to appropriate the individual scientist in order to transcend the idiosyncratic, 'to use an individual's experience as a sampling device for gaining an understanding of the structural and normative'.[29] A similar ambiguity can be found even in Hankins's defence of biography. On the one hand, he maintains that 'letters written under great emotional stress are the best grist for the biographer's mill, because they lead straight to the heart of the subject's personality and reveal the groundsprings from which his actions come', but on the other hand, he endorses the view of the person as the focal point of larger social

---

[24] Hankins (1979), 2.
[25] Theerman (1985).
[26] Young (1988).
[27] Abir-Am (1991), 342.
[28] Keller (1983), xiii.
[29] Rosenberg (1988), 569. Rosenberg asserts that he does not want to 'denigrate biography as a genre or prescribe a particular style of biography' (note 3), but it is nevertheless difficult to read the quoted passage as anything but an acceptance of the subsumation of biography under social or institutional history of science.

factors: '[Biography] gives us a way to tie together the parallel currents of history at the level where the events and ideas occur. . . . We have, in the case of an individual, his scientific, philosophical, social and political ideas wrapped up in a single package', writes Hankins.[30]

This ambiguity has not favoured 'pure biography'.[31] On the contrary, the renaissance of science biography in the 1980s coincides with a largely tacit redefinition of the genre as 'social biography':[32] from being an art of telling individual lives in science to becoming an auxiliary to the social history and sociology of science. This shift of aims may be one of the reasons why Susan Sheets-Pyenson, in spite of so many voices to the contrary, believes that '[h]istorians of science today . . . have scarcely rejected the biographical approach'.[33] A recent example of the tendency to redefine biography in terms of social biography is Smith and Wise's otherwise laudable study of Lord Kelvin, unfortunately subtitled 'a biographical study' in spite of the fact that the authors deliberately chose to write about a person who left very few sources about his personal life.[34] Another, more subtle, example is provided by Desmond and Moore, whose purpose with *Darwin* is to correct the portraits painted by 'textual analysts and historians of disembodied ideas', and to write a biography that follows in the wake of 'the recent upheaval in the history of science, and its new emphasis on the cultural conditioning of knowledge'.[35] Their Darwin is a person plagued by self-doubt, stomach aches and constant worries about his respectability, but he is nevertheless primarily 'a product of his time' and of the social context, and consequently theirs is 'a defiantly social portrait': only by showing Darwin against the background of reform bills, poor law riots, industrial innovation, and so forth, will 'his evolutionary achievements make sense', the authors suggest.[36]

The general trend during the last decade to shift the focus of the genre, from the life of the scientist to social biography, is not just a matter of programmatic statements and the intentions of individual biographers. It is also an effect of the way biographies are read and received. Although readers' responses are notoriously difficult to evaluate, the favourable attitude to social biographies displayed by most reviewers, derive, I believe, from the

---

[30] Hankins (1979), 5.
[31] Kendall (1986), 49.
[32] Eickelman (1985), xv.
[33] Sheets-Pyenson (1990), 399.
[34] Smith and Wise (1989).
[35] Desmond and Moore (1991), xviii–xx.
[36] Ibid.

fact that the genre of science biography today is embedded in a broadly defined sociological discourse – a discourse centered around science and scientists as products of a specific culture and a social and political context. For instance, in an essay review of the 'Darwin industry', Timothy Lenoir, although sensitive to the problem of the formation of Darwin's identity, nevertheless emphasises that it is 'by concentrating on the social matrix' within which Darwin worked out his professional identity that we can gain new insights into his theoretical development.[37] Another reviewer simply suggests that the problem of biography's relevance for 'the new [i.e. social] history of science' can be solved by redefining the genre of biography – instead of focusing on the personality of individual scientists, the historian should use biographical material as a 'convenient indicator of the possibilities for action offered by a particular society'.[38]

The post-structuralist challenge to the biographical subject

The guiding idea of this chapter is that science biography is not just a 'convenient indicator' of social action, but a genre with a clearly defined topic of its own – the individual scientist and his existential project. Throughout my work on the Jerne biography I have been able to draw on rich material, including diaries, private correspondence, and in-depth interviews that provide access to these aspects of his life. I have indeed often been tempted to utilise the papers and the interviews with Jerne and other contemporary immunologists to write the history of recent immunology instead, particularly since as this history has only recently begun to be explored.[39] Jerne was a major actor in the cellular and molecular transmutation of immunology in the 1950s–1970s, as the discipline evolved into 'a subtle and sophisticated science out of the boredom of blind serology'.[40] Through his theoretical and methodological work on the antibody problem – the selective theory of antibody formation, the identification of single antibody producing cells, and the network approach to the immune system – Jerne placed himself at the centre of the disciplinary discourse for almost three decades.

But a temptation to write history of science is not identical with a 'necessity': there is no historiographical 'iron law', not even an unwritten law of

[37] Lenoir (1987), 127.
[38] Morus (1990), 520.
[39] Corbellini (1990), Moulin (1991); Tauber (1994).
[40] Pernis and Augustin (1982), 1.

good writing manners, that demands that life stories must serve simultaneously as biography and as history of science. Whatever rich contextualisation science biographies might provide of the relation between science and society, social biography is only one among several approaches to the understanding of a life in science. Thus, neither the philosophical dismissal of the person, nor the sociological dismissal of the individual has fundamentally shaken the legitimacy of the genre as such. The task of freeing biography from the cognitive reduction of philosophy of science is rather unproblematic – the rigid separation of cognition from passion, mind from body, and reason from imagination has become increasingly difficult to defend, even philosophically.[41] Likewise, the task of freeing the genre from the false necessity of always having to take the social context into consideration is unproblematic, since it only demands the drawing of a clear distinction between seeing life histories in their historical situatedness versus as 'an important subject in its own right' – this involves, as Jerald Wallulis points out, another conception of consciousness: 'The consciousness of having been enabled, as a necessary and useful complement to historically effected consciousness'.[42]

Hence, the problem with the turn towards social biography is not its presence as such, but rather the hegemonic ambitions and derogatory attitudes from the side of some of its promoters towards those who study scientists in their own right – a tendency that has recently provoked L. Pearce Williams to stem the tide of 'the social swamp', as he calls it, with a polemical defence for the right to focus on the uniqueness of the individual scientist. He does so with an argument that can be traced back to the dictum of Thomas Carlyle, that 'history is the essence of innumerable biographies', a position that was systematised by Wilhelm Dilthey in the last century in his foundation of the *Geisteswissenschaften*. 'There is only a society which each individual constructs for him- or herself. . . . Every person, then, at least in part, lives in a different society', says Williams, echoing Dilthey, who saw biography and autobiography as the beginning and the end, respectively, of the human sciences.[43] Williams' position is a healthy antidote to the sociological reflexes that characterise much history of science today, and which may be usefully undisciplined by the biographer's individualism. But his main argument against social historians and sociologists of science ('there are giants,

[41] Johnson (1987).
[42] Wallulis (1990), 134–5.
[43] Williams (1991), 207; Dilthey (1989).

and I think it important to study them')[44] is impotent against the third wave of challenge to biography – the post-structuralist critique of the foundational character of the subject. During my work on Jerne, I have been repeatedly shaken by the prospect of losing the subject of my work, since adherents to late structuralism (post-structuralism) have raised such serious questions about what they claim to be a naive realist conception of the subject that their criticism seems to preclude not only the 'giants' but the genre altogether.

The pivotal element in post-structuralist thinking is the problematic character of the referent: '[I]n the field of the subject, there is no referent', said Roland Barthes.[45] On this view, language is not referential, there is nothing outside the text, and meaning is only produced textually; hence, the referential character of concepts is dissolved into metaphors and figures of speech. The human subject is thought to be no exception: it is 'merely an effect of language',[46] a product of discourse, constituted through language and rhetoric, 'dispersed, divided and decentred by language'.[47] Since persons are textually produced, they are said to lose any given nature, any unitary identity; human beings are simply 'incarnated vocabularies'.[48] The post-structuralist notions of the deconstruction of the subject and the pronouncement of the 'death of the author' have had a widespread influence on the interpretation of autobiograhical texts.[49] Although the genre of biography has been slower in giving up the notion of referentiality, it has been increasingly squeezed between anti-referential autobiography and anti-referential tendencies in historical writing. Critics of literary biography question the 'myth of personal coherence' and emphasise the 'discontinuity' of the self.[50] With reference to the works of Michel Foucault and Jacques Derrida, critical art historians likewise denounce the ontological primacy of the author, reject the traditional genre of art biography with its implicit idea of the artist as essence, and consider biographies to be full of naive and undocumented attempts to relate a particular work of art to the psychological life of its maker.[51]

The step to announcing the disappearance of the subject of science

---

[44] Williams (1991), 204.
[45] Barthes (1977), 56.
[46] Ibid.
[47] Christie and Orton (1988), 556.
[48] Rorty (1989), 88.
[49] For a review of the referentiality problem in autobiography, see Eakin (1992).
[50] Clifford (1978), 44–5.
[51] See, e.g., Cranshaw and Lewis (1989).

biography is a short one. In fact, one sociologist of science recently announced the death of the individual subject in science: 'The physicist as an individual is extinct', 'the epistemic subject is no longer the single scientist', and '[s]ubjectivity seems to be pretty much lost in the process'.[52] Therefore, we can soon expect science biography to be challenged by the same post-structuralist critique that has already haunted literary biography and art biography.[53] In his anti-biography of Edison, David Nye suggests that 'the fundamental error of biography lies in the attempt to construct a definitive figure at all': since individuals are 'divided selves who remain essentially unknowable in their endless variations', their lives cannot be recovered.[54] Nye's correction for the alleged error of biography is to avoid constructing a narrative line in the pursuit of a central consciousness. True, he claims that biographers must take the continuity of the living body and its stream of consciousness for granted – but this does not require a continuity in mind contents, in habits or in behaviour, or in ways of being in science or being an intellectual, he says: 'If [the person] may be expressed as sixteen different figures, [the biographer] will do so rather than perform a reduction'.[55] Not *the* Edison, but a plurality of Edisons. The recipe for science biography from the side of post-structuralism seems to be that Holmes should rather have written 'Hans Krebs: a Proper Name with Sixteen Unfixed Identities', instead of *Hans Krebs: The Formation of a Scientific Life*.[56] Thus, from the ideal vantage point of post-structuralism, the enterprise of science biography seems to be an impossible one. With the substitution of 'Anxieties of Discourse' for 'Portraits of the Artist' the post-structuralist critique seems to undermine any attempt to write science biographies that focus on whole persons.[57]

Yet, a number of arguments can be given against the post-structuralist dismissal of the biographical subject. A pragmatic argument is that biographies are substitutes for the traditional novel and are read the way novels used to be read before literary modernism dispensed with the author and the

---

[52] Karin Knorr-Cetina in a lecture at Program in History of Science, Stanford University, November 1991. Note that this view has much in common with the classical philosophy of science view of science. Scientists were supposed to view science as if the 'author is dead' and to read texts intertextually instead; and they were supposed to treat the history of science from a 'presentist' stance.

[53] For a discussion of a post-structuralist literature biography, see, Clifford (1978); Nadel (1984) and Epstein (1987).

[54] Nye (1983), 17–18, 8.

[55] Ibid., 19.

[56] Holmes (1991).

[57] Bradbury (1992), 8.

subject – hence, the identification of the reader with a real biographical subject constitutes the fundamental motive for the reader's interest in biography.[58] Second, post-structuralists bring to the extreme only one aspect of the
modernist account of our relation to the social context, namely, that our lives
are shaped by social and rhetorical institutions and practices. As Roberto M.
Unger points out, however, this is to focus on one side of modernity only
while neglecting the other side of its grand lesson: 'That we can always break
through all contexts of practical and conceptual activity'.[59] (After all, I suppose that the reason why even post-structuralists remain in academia is the
inherent potentiality that scholarly work gives for self-expression and creation of spaces of their own). A related counter-argument is that post-
structuralists have a metaphysical subject hidden in the closet. To take their
position seriously, one would expect that the deconstruction of the subject
applies symmetrically to themselves. Whenever they talk about their own
work, however, even when they do so reflexively, they talk autobiographically about themselves and their intentions, so that willingly or unwillingly,
reflexively or autobiographically they refer to their own authorial identity.
Yet another related objection is that the deconstruction of the subject is 'para-
sitic', to borrow an expression from Richard Rorty, upon reference to personal identity.[60] Deconstruction is a necessary restraint to our easy habits of
typifying other humans and then believing that the typifications correspond
to reality. But to insist on continuing deconstruction where common sense
or, as I suggest in this article, an existential understanding, would do – 'to
make ourselves unable to view normal discourse in terms of its own motives,
and able to view it only from within our own abnormal discourse' – is 'not
mad, but it does show a lack of education'.[61] And to attempt to deconstruct
the human subject without even recognising the abnormality of the stance
'is madness in the most literal and terrible sense'.[62]

Post-structuralism provides us with an anonymous smorgasbord of
texts – take what you can use for your own purpose, a piece here, a piece
there, and construct your own story from the fragments. In the next moment
the story will again be fragmented, and some of it used to construct new

[58] Cf. Eakin (1992), 36.
[59] Unger (1984), 8. Cf. Giddens (1991): 'The self is not a passive entity, determined by external
influences: in forging their self-identities, no matter how local their specific contexts of action,
individuals contribute to and directly promote social influences that are global in their consequences and implications' (p.2).
[60] Rorty (1980), 365.
[61] Ibid., 366.
[62] Ibid.

texts, and so forth, *ad infinitum*; a vision of intellectual work that corresponds to undisciplined word-processing. In this vision of intellectual life, the question of authorship is indeed superfluous. In contrast, biography brings in not only the author but the *Oeuvre*, the accomplishments of a lifetime. There is, as Alisdair MacIntyre points out, a narrative unity to life.[63] Criticising MacIntyre's thesis of the narrative unity of life, David Cooper believes that its thrust can be saved by casting it in the prescriptive mood: 'If a person's life has not possessed narrative unity, there is only one way, formally speaking, by which he can, without self-deceit, come to view it as if it had. And this is to do something which actually confers narrative unity upon it'.[64] Thus, even if we do not believe in the narrative unity of life, we can believe in the possibility of constructing our lives. I cannot see any great difference, however, between MacIntyre's position and Cooper's advocacy of the Nietzschean demand 'that we so live now and in the future as to confer a telos on the past by our exploitation of it'.[65] Would not MacIntyre be able to say that the Nietzschean demand can be made by everybody, at any time in his life, so that at any point in time, a person can so live as to confer a telos on the past by his exploitation of it? If so, the person's life could be seen as a succession of points in time when he has applied the Nietzschean demand. There is no reason why this succession of points in time could not be made seamless, simply by increasing the number of points in time until they coalesce, and as a consequence one ends up with a continuous narrative lifecourse, which was MacIntyre's original thesis.

Furthermore, the post-structuralist argument for dismissing the unity character of the scientist is compelling, but hard to uphold in the long run when one embarks on writing a biography of a contemporary scientist based on repeated interviews during an extended time period. In the beginning of this work, I was inspired by the idea of a polyphonic biography.[66] Each interview with Jerne, and particularly with his friends and colleagues, gave rise to a slightly different 'Niels Jerne'. Not even narratives of the same historical event were identical, since each informant gave a new story about the person and his acheivements. Sixteen accounts would simply not do it. Obviously, 'Niels Jerne' varied, both with the context of earlier interactions between him and the informant and with the context of the

---

[63] MacIntyre (1982).
[64] Cooper (1988). 168.
[65] Ibid., 170.
[66] Particularly by Eickelman (1985).

interaction between me and the informant during the interview. This experience first led me to a view of biography similar to that of the post-structuralist vision. But with an increasing number of interviews with Jerne, after 50–100 hours of discussion, the notions of seemingly endless variability of texts, the 'death of the author', and the non-referentiality of the proper name became increasingly absurd. I came to the conclusion that it is only occasional acquaintances that can be seen as intertextually constituted. In other words, as the result of a long period of interaction the biographical subject turns from an 'it' to a 'thou'.[67] Through empathic engagement in a series of interviews, and through the repetitive character of the interaction itself, the abstract character of the biographical subject becomes increasingly concrete. As Jamie Ferreira writes with reference to Søren Kierkegaard's notion of 'repetition': 'A concrete being is, because free, irreducible; the irreducible requires endless exploration, eternal re-seeking, endless demanding back. Thus, the inexhaustibility and complexity of the concrete are what both allow and require repetition in order that justice be done to the concrete'.[68]

Finally, even if we should for a moment accept the idea that lives are linguistically and socially constituted and that a biographer cannot refer to any foundational self for choosing one biographical narrative over the other, we do nevertheless, in practice, choose some narratives and vocabularies over others. Rorty, who is otherwise an advocate for the linguistic constitution of the subject, argues that we should choose vocabularies that increase human solidarity and heighten our awareness of human suffering.[69] Rorty's hero, the liberal ironist, does not believe in any foundational common truth or common goal that binds humans together, and therefore rejects the classical humanistic conception of a human essence. But he believes that we share a common selfish hope – that our own understanding of the world will not be destroyed and humiliated by others. By reflecting upon the pain so inflicted, we will not arrive at a reason for caring about the other, says Rorty, but we can nevertheless make sure that we notice suffering when it occurs.[70] However, is not this occurrence of suffering and pain precisely an essential conception of the subject? How should we be able to notice and heighten our awareness of it unless we take for granted that it is

---

[67] Buber (1958).
[68] Ferreira (1989), 22.
[69] Rorty (1989).
[70] Ibid., 93.

a universal human trait? Evidently, human solidarity demands that we treat at least the suffering human subject as having referential reality.

An existential approach to science biography

In Rorty's vision the imaginative ability 'to see strange people as fellow sufferers' is not a task for (literary) theory, but for genres such as reportage, docudrama, theatre, movies, and particularly the novel.[71] To this list I would add biography: I suggest that we can overcome 'the fundamental error of biography' by means of existential biography. So far, however, existential thinking has made very little impact on the history of science or on the art of science biography, compared with the impact of psychobiographical approaches, including psychoanalytical thinking.[72] Given the character of scientific research as compared with many other human activities, including its creative aspects, the elements of high risk enterprise, and its often transcendental character,[73] this lack of interest in the existential approach among historians of science in general and science biographers in particular is remarkable. A notable exception is Maila Walter with her biography of Percy Bridgman. While also conveying the sensibilities of the time (as reflected in its title), *Science and Cultural Crisis* is primarily 'a story about the meaning of science – its meaning for an individual in a particular culture in a particular era'.[74] Walter focuses on the existential grounds of scientific truth and demonstrates in detail how Bridgman's physics, and his later particularist philosophical outlook, was thoroughly embedded in his personal struggle. The portrait reflects its subject's own view of science: 'The checking and judging and accepting that together constitute [scientific] understanding, are done by me and can be done for me by no one else. They are as private as my toothache and without them science is dead', wrote Bridgman in his philosophical reflections.[75] Other exemplary biographies from the point of view of an existential approach to intellectual lives are Ray Monk's portrait of Ludwig Wittgenstein, with its focus on the

---

[71] Ibid., xvi.
[72] The situation is not much different for biography as a genre in general. In his overview of psychobiography, Runyan (1984) mentions a number of alternatives to psychoanalytic psychobiography, including humanistic and existentialist biography, but says nothing about existential biography beyond references to Sartre's biographies.
[73] Forman (1991).
[74] Walter (1990), 1.
[75] Quoted in Walter (1990), 170.

philosopher's search for a purity in thought he realised he would never be able to achieve,[76] and James Miller's recent study of Michel Foucault, which approaches the philosopher's writing 'as if it expressed a powerful desire to realise a certain form of life'.[77]

An existential approach does not mean a rejection of the importance of the social life of the individual, nor does it involve an uncritical individualist viewpoint. The relation between the individual and society – the contrast between the life of an autonomous and authentic individual and the life of a public individual immersed in society – is a persistent theme in twentieth-century existentialist writings.[78] On the one hand, the individual has been seen as a participant in, or even a product of, a public, social world, and formed by the 'Look', by the judgements and categorisations by others. But this characterisation of a life does not tell the whole story about a human existence. The decisive point is that even if a large portion of every human life is lived inauthentically under the spell of others, human beings have the capacity to undo this condition – authenticity can be won in struggling out of an everyday condition of inauthenticity – a conclusion which is particularly significant in the sphere of human activities called science. One can, and most scholars in science studies in the last decade have done so, emphasise the communal aspects of science, how the members of scientific 'disciplines' are objects formed by the 'Look of the paradigm' or the 'episteme'. But one can also emphasise the constant efforts that scientists make to break out, their disobedience to the rules and discipline of the discipline, and their attempts to retain what Karl Jaspers called the 'original potential': 'Although my social I is . . . imposed upon me, I can still put up an inner resistance to it. . . . Although I am in my social I at each moment, I no longer coincide with it. . . . I am not a result of social configurations . . . [for] I retain my own original potential'.[79]

To what extent is existential biography different from the well-established biographical subgenre of psychobiography?[80] Several science biographers have attempted to apply a coherent psychological theory, including psychoanalytic theory and developmental psychology.[81] There are many good reasons for doing so: scientific psychology in different guises is so integrated

---

[76] Monk (1990).
[77] Miller (1993), 5.
[78] Cooper (1990), particularly Ch. 7.
[79] Quoted in Cooper (1990), 110.
[80] Runyan (1984) and Runyan (ed.) (1988).
[81] E.g., Manuel (1968), Sokal (1990).

in our culture that it is hard for a biographer to avoid incorporating elements of it in narratives of lives of scientists. Psychological models are obviously also of great use in biographical case-studies, and for generalisations about scientific reasoning, creativity, the life-course and so forth.[82] Nevertheless, a certain amount of precaution is to be recommended. The genre of biography is not primarily a generalising, explanatory science, nor is it a critical inquiry by which the subject is analysed with detachment and scepticism and quoted to illustrate some general sociological, philosophical, or psychological principles — it is primarily a genre through which we try to bring to life again the unique individual: 'The life itself is the achievement; not the explanation of it'.[83] Or as Miller says, quoting his subject, the aim with the Foucault biography has not been to conjure up the deep psychological subject, 'but rather the one who says "I" in the works, the letters, the drafts, the sketches, the personal secrets'.[84] Furthermore, existential science biography is ultimately also an analysis of the life of the concrete, individual researcher, not a case-study of what it means to be a scientist in general. For Kierkegaard, the analysis of man is not an abstract investigation into 'humanness' in general, but an analysis of the factual, concrete human life as actually lived. In that sense existential biography transcends the generalising demands of social history and psychobiography.

The notion of existential biography has mainly been used in connection with Jean-Paul Sartre's biographical works, particularly the biographies of Flaubert, Baudelaire, and Genet.[85] Yet, these biographies are problematic as models for the notion of existential biography developed here, on the ground that Sartre's ambition is to redescribe the lives of his subjects into his own mixture of existentialist, Freudian, and Marxist ideas, and thus has difficulties distinguishing the life of the subject from his own literary ambitions.[86] In fact, biographical redescription is an old device in the arsenal of knowledge–power discourses:[87] psychoanalytical thinking, for example, has repeatedly been criticised on this ground. As Rorty points out, 'most people do not want to be redescribed, they want to be taken on their own terms — taken seriously just as they are and just as they talk'. To redescribe people's experiences in

---

[82] See, e.g., Gruber (1981).
[83] Skidelsky (1987), 1250.
[84] Miller (1993), 5.
[85] Scriven (1984).
[86] Shapiro (1986), 357.
[87] Söderqvist (1991).

other terms is 'potentially very cruel'.[88] The science biographer who applies a psychological theory to his subject, threatens the scientist's final vocabulary and ability to make sense of himself in his own terms rather than the biographer's, and thereby suggests that the scientist's self and his world 'are futile, obsolete, *powerless*'.[89] I suggest that the degree of redescription is one of the criteria that distinguishes existential biography from social biography and psychobiography.

The biographer cannot, of course, avoid seeing the scientist as a social being, or drawing psychobiographical conclusions altogether. A certain amount of hermeneutical distance in necessary – the biographer must try to make the experiences of the scientist comprehensible in terms of his own historical, sociological or psychological training, and compare the experiences of the scientist with those of other individuals using other vocabularies about self and the world. The biographer's very task involves a certain amount of redescription. In addition to the necessary distance, however, the biographer must, in one way or the other, adopt an empathetic stance which does not falsify the scientist's position by imposing an alien vocabulary.[90] Biographers who try to respect the subject as a human being must, as far as possible, be sensitive to the vocabulary the person uses about himself, his work and the world around him. Accordingly, the biographer is free to compare, contrast, and challenge the vocabulary of his subject with other vocabularies of his choice – but not to redescribe the scientist in terms of these other vocabularies alone. To a certain extent biographers must then, to borrow a common notion from contemporary anthropology, 'go native'. Since the greatest demand the writing of biography makes 'is an initial respect for the subject as a human being',[91] scientific psychology is therefore not necessarily the most obvious choice for a biographer who focuses on scientists in their own rights, unless, of course, the scientist in question experiences, understands, and describes himself in terms of some scientific psychology (something not even B.F. Skinner was able to do in his autobiography, however).[92] As a consequence, psychological approaches, for example, cognitive psychology, which is otherwise an excellent tool for case-study work, should be applied with care when the aim of the work is to understand the

---

[88] Rorty (1989), 89.
[89] Ibid., 89–90.
[90] Frank (1985).
[91] Anderson (1981), 403
[92] Skinner (1984).

richness and fullness of the life of a concrete individual and his experiences, and to stimulate an 'awareness of the worth of the subject as a human being' as a means for edification.[93]

Different scientists use different vocabularies when trying to make sense of their lives in interviews and autobiographical writings. In principle, therefore, each biographical interpretation has to be based on the personally unique vocabulary of the subject. Let me illustrate this with reference to the biography of Niels Jerne. Throughout his life, in letters and diaries, in our conversations, and in biographical interviews, Jerne often used a vocabulary about the self and the world that incorporates elements from classical authors such as Shakespeare, from nineteenth-century romantic philosophers such as Kierkegaard and Friedrich Nietzsche, and from modernist authors such as Fyodor Dostoyevsky, Franz Kafka, and Marcel Proust. Kierkegaard's ideas have been particularly prominent in his understanding of self and other human beings. Jerne himself says that he discovered Kierkegaard during his high-school years in Holland: 'I believe I was sixteen at the time . . . I found him in my father's library, and I could read him in Danish',[94] and adds that he felt intellectually related with Kierkegaard, a person 'who is like me': 'Oh, he has impressed me, because he writes with courage, with intelligence, a merry mind — undescribable. . . . I like the whole thing. He is so funny, you can laugh, and at the same time so deep. There is so much resonance, like when listening to Mozart'.[95] Several of his colleagues and friends have borne witness to his passion for the great Danish existential philosopher: 'He was drawn to Kierkegaard like a magnet', says a visitor to the Danish State Serum Institute in the late 1940s, 'because some of his longings and perhaps also some of his experiences, tragic experiences of life, made him understand what Kierkegaard's deepest concerns are'.[96]

Jerne's use of Kierkegaard's language emphasises, more than many other ways of speaking about the self and the world, the existential and passionate dimensions of life. It carries a vocabulary of vulnerability and doubt, anxiety and existential loneliness, with little of the pragmatic, energetic jargon of so many biographies and autobiographies of scientists. It has, of course, been

---

[93] Anderson (1981), 403.

[94] 'Ich glaube ich war sechzehn damals . . . Ich fand ihn in der Bibliothek meines Vaters, und ich konnte ihn auf danisch lesen'. Anon. (1985), 8.

[95] 'der ist so wie ich': 'Oh, er hat mir imponiert, denn er schreibt mit einer Wucht, mit einer Intelligenz, einem Frohsinn — unbeschreiblich . . . Es ist das ganze, was mir gefällt. Er ist so lustig, man kann lachen, und gleichzeitig so tiefsinnig. Es gibt so einen Nachklang, wie wenn man Mozart hört.' Anon. (1985), 9.

[96] Interview with Hans Noll by Thomas Söderqvist, September 12, 1989.

modified by later readings and life experiences, and it has increasingly been replaced by a technical immunological vocabulary, but the existential and passionate way of speaking is identifiable also in the scientific correspondence. In an autobiographical essay Jerne even suggests that 'reverberations of Kierkegaard' may have 'contributed to the idea of a selective mechanism of antibody formation',[97] a case analogous to that suggested for the relation between Kierkegaard and Niels Bohr.[98] Accordingly, I have chosen in my own biographical work to utilise an existential vocabulary that resonates with and magnifies Jerne's own understanding of self and others. Throughout this work, I have also come to believe that the value of an existential approach is by no means limited to this particular biographical work. It may be that many scientists have not had Jerne's explicit recourse to a modernist and existential vocabulary, but I am convinced that this approach might provide a language that resonates with the experiences of other scientists as well – provided, of course, that the archive or the interview transcripts contain the necessary source material.

Passion and existential projects in science

I concluded above that human solidarity demands that we treat the suffering human subject as having referential reality as a unitary person. But why stop at suffering? By noting and reflecting on the passions of the scientist – both negative emotions, such as anguish and anxiety, despair and dread, embarrassment and fear, frustration and sadness, and positive emotions such as joy, hope and love – we will be able to transcend the idea of the scientist as a mere 'convenient indicator', and become aware of him instead as a discrete, embodied mind. In contrast to its central position in biographies of artists and authors, however, the topic of passion is not a matter of course in science biography. In fact, the widespread use of the term 'scientific biography' implies a focus on the intellectual and cognitive aspects of the lives of scientists, and a peripheral treatment of the passions. Usually restricted to 'a passion to know', as a collection of journalistic essays on scientists is titled,[99] the passions have so far been marginally treated in science studies.[100] The

---

[97] Jerne (1966), 301.
[98] Jammer (1966), 1040ff.
[99] Hammond (1984).
[100] A few psychologists have dealt empirically with the emotional life of the scientist and of scientific work, e.g. Eiduson (1962). The affective relations in science are also central to Lorraine Daston's the 'moral economy of science'. See Daston (1995). Otherwise the present interest in passions in

sociological turn in history of science has not remedied this traditional neglect of the passionate aspects of science: scientific knowledge is socially,
linguistically, and rhetorically contextualised, but rarely seen as having anything to do with the passions of the scientist, an attitude that has also spilled
over to social biography.[101]

It has sometimes been claimed that passions are socially constructed
too;[102] on the view taken here, however, the passions of the scientist are not
social products but integral elements in the realisation of existential projects,
defined here as the individual's view of how to live in a way that gives a
measure of sense, unity, and value to his life.[103] In our struggles to overcome
the threats of humiliation, suffering, anxiety and pain, and in our hopes of
being able to join with others, we invoke widely different existential projects.
Rather than offering social visions, these projects offer guidance to our life-
courses, particularly to our strivings for empowerment. In Unger's combined
modernist and Christian-Romantic account of personality, empowerment
amounts to the successful diminishing of the conflict between the conditions
that enable us to assert ourselves as persons: on the one hand, our desire to
engage with other people and through this engagement to establish ourselves in the world; on the other hand, our need to prevent this engagement
from subjugating and depersonalising us.[104] The passions embody the realisation of the tension between the conditions for self-assertion: fear, despair,
vanity, pride, jealousy and envy are the results of a failure to achieve
empowerment; hope, faith and love are expressions of our success in this
respect.[105]

From the point of view of existential biography, this ability to handle the
enabling conditions of self-assertion lies at the heart of the life and work of
every scientist. In our attempts to assert ourselves through scientific and
scholarly work, we are permanently at risk. In projecting our existential projects into the social space, in acts of 'world-making',[106] we are constantly at
the peril of being rejected and overwhelmed by others. Scientists who choose

science is negligible – as reflected by a recent textbook on the psychology of science (Gholson *et al.*, 1989) which mentions passion and emotion only in passing.

[101] Even though Desmond and Moore (1991) treat Darwin's emotions at length, 'his fears and foibles'
are said to make sense primarily against the background of activities such as economic invest-
ments and living a squire's life. If one disregards their introductory programmatic professions to
the cultural conditioning of knowledge, however, their Darwin portrait in the bulk text takes on
a much more existential character.
[102] Harré (1986).
[103] Unger (1984), 47ff.
[104] Ibid, particularly 115ff.
[105] Unger (1984).
[106] Goodman (1978).

to go their own way are committed to acts of courage, 'always risking a fearful penalty if they are wrong'.[107] In autobiographical reports several scientist
have used a varied passionate lexicon, for example, the intense feelings of
pain associated with trying to solve a problem, the joy when the solution
comes, and at the same time the feeling of fear, anxiety, even terror during
the process. Despite its seemingly collective nature, science is one of the
most lonely activities in the modern world, and it is often a painful one as
well. The pain may have its origin in activities outside the walls of the laboratory, or it may have its roots in the despairs within. But whatever its source,
pain colours and runs through the life of the scientist, irrespective of his
scholarly standing. In her collection of short biographies of woman scientists, Joan Dash points out that the passions of scientists are strong and per-
vade their whole existence. Scientific research has its drudgeries and long
stretches of boredom and routine, 'yet it seems universal among those
engaged in original research, from the merest postdoctoral fellow to men of
Nobel caliber, that they tend to describe their feelings about their work in
such vivid terms that everything else in life — everything — sounds pale beside
it'.[108] As one scientist says:

> You go through this long, hard period of filling yourself up with as
> much information as you can. You just sort of feel it all rumbling
> around inside of you. . . . Then . . . you begin to feel a solution, a
> resolution, bubbling up to your consciousness. At the same time
> you begin to get very excited, tremendously elated — pervaded by
> a fantastic sense of joy. . . . But there's an aspect of terror too in
> these moments of creativity. . . . Being shaken out from your
> normal experience enhances your awareness of mortality. . . . It's
> like throwing up when you're sick.[109]

Similar passions and bodily sensations go through many autobiographical
narratives. In somewhat less dramatic terms Jerne describes how he felt the
weeks before he formulated the somatic generation theory of antibody diversity:[110]

> [In early July of 1969] I was hit by a spell of creativity that lasted
> until the day before yesterday. Being aware, I followed my own
> behaviour quite carefully; I felt that all the chores (such as farewell

---

[107] Goodfield (1981), 235.
[108] Dash (1973), 279.
[109] Quoted in Dash (1973), 318.
[110] Jerne (1966).

> speeches in Frankfurt, etc) were merely nothingness. I had the feel-
> ing that I had a good idea somewhere though I did not quite under-
> stand what it was. Fact is, that I was very nervous, stopped eating,
> writing, etc. until 20 July. Like a log coming slowly to the surface
> of a lake, I knew what I wanted to understand. It is now laid down
> in the attached manuscript that I got finished a few days ago.[111]

Likewise Paul Dirac speaks about the 'feelings of a research worker when he
is hot on the trail and has hopes of attaining some important result which
will have a profound influence'.[112] He is filled with hopes and fears: 'I don't
suppose one can ever have great hopes without their being combined with
great fears'.[113] With specific reference to H.A. Lorentz's 'near miss' of the
theory of relativity, Dirac discusses how fear can hold a scientist back from
completing his work: 'He did all the hard work – all the really necessary
mathematics', Dirac say, 'but he wasn't able to go beyond that and you will
ask yourself, why?':

> I think he must have been held back by fears. Some kind of
> inhibition. He was really afraid to venture into entirely new
> ground, to question ideas which had been accepted from time
> immemorial. He preferred to stay on the solid ground of mathemat-
> ics. So long as he stayed there his position was unassailable. If he
> had gone further he wouldn't have known what criticism he might
> have run into. It was the desire to stay on perfectly safe ground
> which I presume was dominating him.[114]

The point here is not whether Dirac was right in his interpretation of Lorentz
or not, but the fact that he identified passions, such as fear, as an important
element in scientific work.[115] Other biographical and autobiographical portraits remind us that to the scientist, perhaps no fear is stronger than that
which Harold Bloom calls the 'anxiety of influence', the 'horror of finding

---

[111] Jerne to Günther Stent, August 8, 1969 (Jerne papers, Royal Library, Copenhagen).
[112] Quoted in Dresden (1988), 462.
[113] Ibid.
[114] Ibid.
[115] Dirac's awareness of the role of passion in science seems to be contradicted by Kragh, who in
summarising the wealth of anecdotes circulating about the austere and shy physicist, concludes
that theoretical physics was for Dirac 'a substitute for human emotions' (Kragh 1990, 255). However, this biographical portrait has been questioned by Dresden, who, having known Dirac personally, rather remembers him as a 'deeply compassionate human being . . . with concerns, hopes,
fears, and ambitions', and therefore repudiates Kragh's portrait for 'its lack of passion' (Dresden
1990).

himself to be only a copy or a replica'.[116] It is not only the fear that one's works will be forgotten or ignored, but also that, 'even if they are preserved and noticed, nobody will find anything distinctive in them',[117] that they will be redescribed in terms of other findings, or reduced to replicas.

Max Dresden's biography of Hendrik Kramers provides another example of the pervasive impact of fear in a scientist's life and work.[118] As a university professor in the 1920s, Kramers was supposed to live 'in a world of pure reason, a world where there is no fear, anxiety, inadequacy, anger, or passion'.[119] Kramers knew that it was a caricature, and privately he frequently expressed frustration: 'He was more often torn by doubts and beset by fears, which often guided him in paths which led nowhere. Fear and anxiety about his role in physics were his constant companions'.[120] Did these passions have cognitive implications as well? Lewis Feuer suggests that scientists look for conceptual worlds that will answer to their 'emotional longings' and that established theories are 'isomorphic' with the world 'emotionally sought': '[W]e must necessarily enter upon biographical and psychological considerations to ascertain what indeed were the basic emotional longings of the scientist, what the kind of world it was that he, on emotional grounds, sought to realise in his scientific theorising?'[121] What, in other words, asks Feuer, was the scientist's *emotional a priori?*[122] In the Kramers case, Dresden demonstrates that the Dutch physicist was continuously plagued by doubts and concerns about the shortcomings of his accomplishments and returning feelings of fear and uncertainty. He 'expected to mold [the development of his science] and guide it along lines consistent with his views. The resulting struggles, disappointments, successes, heartbreaks, frequently missed opportunities, and rare moments of elation — all these are now hardly remembered. Yet it is only through a detailed understanding of these conflicts and struggles that a genuine appreciation of the significance of the advances can be obtained',[123] claims Dresden, and he identifies a connection between Kramers' difficulties in committing himself to his wife and his difficulties in committing himself to physics: 'There is a striking similarity between Kramers' unwillingness or inability to commit himself to physics in

[116] Bloom (1973), 80.
[117] Rorty (1989), 23–4.
[118] Dresden (1988).
[119] Ibid., 444.
[120] Ibid., 486.
[121] Feuer (1978), 378, 380.
[122] Ibid., 402.
[123] Dresden (1988), 7.

his student years, thereby giving up all other intellectual pursuits, and his indecision in his relation to Storm [his wife] – which would similarly involve a commitment, with a corresponding renunciation of other options'.[124]

Hope is also an 'emotional a priori'. In a study of Ilya Metchnikoff, the founder of the phagocytosis theory of immunity, Alfred Tauber and Leon Chernyak speculate about the relation between Metchnikoff's personality disturbances and his theoretical achievements.[125] During the early years of his career, the Russian zoologist not only expressed a pessimistic *Weltan-schauung* and a belief in the disharmony in Nature but also had physical problems combined with a depressive character. Tauber and Chernyak argue that Metchnikoff's research concerned with the problem of harmony (organismic integrity) resulted both in elaborating a pessimistic personal philosophy and the tragic existential posture that led him to suicide attempts. The turning point was 'the hope . . . that he might solve the problem of integrity' which 'changed not only the direction and the field of his scientific occupation but also his philosophical ideas and apparently deeply altered his personality'.[126] The role of 'hope' in Metchnikoff's life and science points, like the role of 'fear' in Dirac's and Kramers' lives, to the centrality of the notions of passion and existential project in the genre of science biography.

Existential choice in science

In this section I will develop some preliminary remarks about another central topic in an existentially oriented science biography, namely the problem of existential choice. The connection between life and work is a classic problem in science biography, and several attempts have been made either to solve or avoid it. From the point of view of the existential approach to science biography discussed above, life and work are necessarily inseparable: theoretical thinking, experimental design, empirical observation, writing a paper and

---

[124] Ibid., 116. Paul Forman (1991) has recently accused Dresden for falling prey of attachment to a quasi-religious sentiment of 'transcendence' in science, for having celebratory intents and for expressing 'metaphors of religious transcendence, salvation and saintliness' (p. 76) – in short for being whiggish. Scientists should refrain from trying to write biographies of (other) scientists, says Forman, unless they can stay clear of the transcendent sentiment. But are scientists really that handicapped? Even though approving of Dresden's 'considerable sensitivity, even courage, in dealing with the personality and personal relations of his subject', Forman does not acknowledge that this sensitivity to Kramers' existential predicaments probably stems from the author's transcendent attitude and experiences of personal involvement in scientific work. It is presumably easier for someone who has felt the hopes and anxieties of scientific work in his own mind and body to understand the existential dimension of other scientists.

[125] Tauber and Chernyak (1991).

[126] Ibid., 176.

participating in a meeting are integral parts of existential projects, that is, visions that guide our strivings for empowerment. A biography is existential also if it expresses the dilemmas of a person who, trying to assert himself through creative work, has to deal with the fundamental choices of his existence. Contrary to the idea that the scientist is socially constructed, or 'a product of his time', the point of departure for the existential approach is therefore to understand the scientist as he is confronted with his freedom, with his anxiety as he fathoms the consequences of his choices, and, having made the choice, with his feelings of guilt.

To Polanyi rational knowing involves an existential participation of the knower: 'The shaping of knowledge is achieved by pouring ourselves into new forms of existence',[127] and he describes 'the tacit dimension' of scientific discovery as involving 'existential choice':

> We start the pursuit of discovery by pouring ourselves into the subsidiary elements of a problem and we continue to spill ourselves into further clues as we advance further, so that we arrive at discovery fully committed to it as an aspect of reality. These choices create in us a new existence, which challenges others to transform themselves in its image. To this extent, then, 'existence precedes essence', that is, it comes before the truth that we establish and make our own.[128]

In her story of a discovery based on interviews with the rank and file life scientist 'Anna', June Goodfield implicitly draws on an existential understanding of the scientist, and points out that science involves a series of choices. It is first expressed in the very act of deciding to become a scientist at all, then in the 'choice of the particular road one goes down', or in 'choosing not to go down it at all'.[129] The initial strategy of the scientist's experiments may be socially determined; still it is the individual scientist who chooses to reject the preselected strategy and strike out on her own. The choice can be trivial, as whether to choose to work in the lab or in the archive, but it can have severe consequences as well, as in the case of Bridgman, who, by adopting the standpoint that science is essentially private, 'not only gave up the comfort offered by the warmth of community, but the possibility of certainty as well. He was alone is an indifferent world'.[130]

---

[127] Polanyi (1959), 34.
[128] Polanyi (1966), 80.
[129] Goodfield (1981), 234.
[130] Walter (1990), 174.

There is a more fundamental meaning of the notion of choice involved here, however. The notion of existential choice was defined by Kierkegaard in his discussion of the choice between the aesthetical and ethical in *Either/Or*.[131] The 'either-or' dichotomy does not denote a choice between this or that action, or even between good and evil, but a choice between an ethical life, that is, to live a life where questions of good and evil guide your actions, or to remain in an aesthetic life stage, where questions of good and evil are basically irrelevant.[132] Kierkegaard in fact operates with two aesthetic modes: on the one hand the sensuous and immediate aesthete, as exemplified by Don Giovanni in Mozart's opera, and the reflective and abstract aesthete, for example, Johannes in *The Diary of the Seducer*, on the other.[133] By implication scientists are reflective aesthetes, as illustrated by the autobiography of the molecular biologist James D. Watson.[134] The leap into an ethical stage does not imply that the scientist abandons science, but that aesthetic priorities (search for beauty, or truth, or doing research for the sake of joy and so forth) become subordinated to ethical priorities.

What would an ethical life in science look like in contrast? '[T]he ethical individual is transparent to himself', says Kierkegaard, 'and does not live *ins Blaue hinein* as does the aesthetical individual'. This makes the whole difference. He who lives ethically has seen himself, 'knows himself, penetrates with his consciousness his whole concretion, does not allow indefinite thoughts to potter about within him, nor tempting possibilities to distract him with their jugglery; he's not like a witch's letter from which one sense can be got now and then another, depending upon how one turns it. He knows himself'.[135] However, *gnothi seauthon* is not enough to characterise an ethical life according to Kierkegaard: 'The ethical individual knows himself, but this knowledge is not a mere contemplation (for with that the individual is determined by his necessity), it is a reflection upon himself which itself is an action, and therefore I have deliberately preferred to use the expression ''choose oneself'' instead of ''know oneself''.'[136]

Thus, how one scientist may choose himself as an ethical individual, how this choice penetrates his scientific work and achievements, and, conversely,

[131] Kierkegaard (1946); Kierkegaard points back to Socrates as the paradigmatic 'ethical' individual.
[132] Ibid., 143ff.
[133] Ibid. The distinction between the sensuous and the reflective aesthete is not so obvious *in Either/Or*, but becomes clear in *Concluding Unscientific Postscript* (Kierkegaard, 1992). I am grateful to Timothy P. Jackson for pointing this out.
[134] Watson (1968).
[135] Kierkegaard (1946), 216.
[136] Ibid.

how another scientist may refrain from choosing himself, and thus remain in the despair of an aesthetic life – these are the core issues in an existential biography. An existential reconstruction of the subject's life is therefore made from the inside, in an attempt to narrate the development of his life 'as it is directly experienced by the biographical subject'.[137] Hence, existential biography is distinct from (a) social biography, in which the individual is contextualised with reference to his 'situatedness' in a certain time, a certain culture, etc; (b) psychobiography, in which certain traits of the subject's personality or his achievements are explained with reference to psychological theory; and (c) biographical case histories aimed to generalise about genius, creativity, or the life cycle. All such approaches are external to the experiencing individual confronted with his existential choices.

The focus on subjective experience can be further qualified by Kierkegaard's distinction between 'inward' and 'outward' life history. Outward life history is the story of the strife through which the individual tries to acquire something, the strife in which he overcomes the hindrances to possess something. The conscientious biographer, who tries to understand the significant events in the subject's life, and the scientist-as-autobiographer both describe the life in the same way as Kierkegaard's reflective aesthete, the author or the poet, who relates an individual life as it was concentrated in the moment:

> Imagine, then, a knight who has slain five wild boars, four dragons,
> and delivered three enchanted princes, brothers of the princess
> whom he worships. In the romantic chain of reasoning this has com-
> plete reality. To the artist and the poet, however, it is of no import-
> ance at all whether there are five or only four monsters slain. . . .
> He hastens on to the moment. He perhaps reduces the number, con-
> centrates the toils and dangers with poetic intensity, and hastens
> on to the moment, the moment of possession. To him the whole his-
> torical succession is of comparatively little importance.[138]

Kierkegaard's description of the aesthete's concentration on the 'moment of possession' is similar to the way most scientific lives are written. Neither biographers nor scientist-autobiographers see any point dragging in all the details. The daily routines do not really matter. What matters for them is the significant moment – the moment of discovery, the moment when a new model of Nature was conquered and possessed.

[137] Scriven (1984), 45.
[138] Kierkegaard (1946), 112–13.

When it comes to inward history, however, the life of the individual cannot be concentrated in one, or a few, single moments, since it deals with the succession of a life in time where 'every little moment is of the outmost importance'.[139] Whereas pride can be represented in the outward history ('for the essential point in pride is not succession, but intensity in the moment'), humility cannot, 'because here if anywhere we are dealing with succession'.[140] Romantic love can be represented in the moment, but not conjugal love, 'because an ideal husband is not one who is such once in his life but one who every day is such'.[141] The same goes for courage versus patience, the hero versus the cross-bearer. Accordingly, the biographer who focuses on the significant events of the life of the scientist, and the scientific aesthete who in his autobiographical concentration deals with the moment, both have difficulties in catching the non-dramatic succession of all the small events in the ethical life of the individual scientist – the many times when he cared for a graduate student, the seemingly infinite number of times when he waited patiently for the experimental results to come, and the humility with which he accepted contradictory data. Nothing of this lends itself to a dramatic biographical narrative, and yet it is an essential aspect of what it means to lead an ethical life in science. In the annals of history it is outward contributions that counts, whereas in the ethical life of the individual the inward succession of everyday life means everything. To grasp the inwardness and yet write a biography that anybody cares to read and can read intelligibly – this is the challenge to an existential approach to science biography.

Science biography as an edifying genre

Let me in conclusion return to the question that opened this chapter. What is the aim of science biography? The last two decades of social history and sociologically informed history of science have demonstrated beyond any doubt the 'contextual' nature of science – in the context of society at large, in the context of social and political institutions, in the context of gender, class and race, and in the context of the local, contingent settings of the scientific laboratory or in the field. Few would deny the permanent value of this research programme. But is is important to remember that much of this

---

[139] Ibid., 113.
[140] Ibid.
[141] Ibid., 114.

history and sociology of science is driven by what Paul Ricoeur calls a 'hermeneutics of suspicion'.[142] The demonstration of the social and political context of science and the socially constructed character of scientific knowledge has often been seen as ways to deconstruct 'naive realism', to dispel illusions about the power structures operating within and behind science, and to lay bare a naive scientistic ideology of a value-free and context-independent 'search for truth'. An illustrative example is Bruno Latour's *Science in Action*, the explicit normative agenda of which is to help individual scientists by exposing how the machinery of 'Technoscience' works and 'to provide a breathing space to those who want to study independently the extensions of all these networks'.[143]

It should not be denied that this joint sociological and historical discourse constitutes a healthy antidote against traditional history of science and science biography with its authors' often seemingly mindless focus on the chronology of events and achievements of the individual scientist, and strong tendencies towards uncritical hero worship. My only caveat is this: whatever working according to this agenda may expose, it leaves the individual scientist defenseless against the very same powers its promoters want to disclose. Exposure strategies have rarely produced any 'breathing spaces', as the general failure of one ot these, namely Marxism, has amply demonstrated. Individual scientists may become more disillusioned, and probably also more cynical, after reading studies that demonstrate the constructed character of knowledge and one-sidedly focus on the social and political context of science. But they hardly become more able to resist 'technoscience'. For that reason scientists do not need more historical or sociological studies of the system of science to acquire breathing spaces — they need to strengthen their personal ability to breathe.

This is where existential biography enters the picture. The basic argument of this chapter is that the aim of science biography is not primarily to be a genre that adds yet another means for disclosing the contextual and socially constructed nature of science. Its primary aim is to be a genre which conveys an understanding of what it means to live a life in which scientific work and rational thinking are part of an existential project and involves existential choices. The aim of existential biography is to help scientists and non-scientists alike to strengthen their abilities to live fuller and more authentic

---

[142] Ricoeur (1970).
[143] Latour (1987), 257.

intellectual lives.[144] Instead of adding to the hermeneutics of suspicion that governs so much of today's history and sociology of science, science biography should rather contribute to a 'second naívité', a 'hermeneutics of belief', based on trust and a willingness to listen in order to understand.[145] The ironic relationship imposed between the biographer, his subject and the reader is denied to the existential biographer. Similarly, the reader is asked primarily to identify with the biographical subject rather than just contemplate his plight or withdraw into judgement.[146]

An existential approach to biography points to a dimension of uniqueness and individual choice. To stress the notion of the existential project and the notion of empowerment that goes with it is not to deny the importance of the social and political contexts of our actions. But those who stress the notion of social context neglect the other side of the modernist coin, that is, that we *are* able to break through these contexts. To give attention to our abilities to break contexts is to give the freely acting, ethically responsible, individual scientist the privileged role in science biography. Contrary to Clark Elliott, who grants the value of biography for understanding pre-World War II science as one of several approaches to history of science, but questions whether organised Big Science and team research in post-World War II science 'leave a legitimate place for the study of individual lives',[147] the guiding normative idea of this chapter is that it is precisely the emergence of Big Science, collective team research, and anonymous technoscience structures that makes it urgent to focus on the lives of individual scientists without the constant, often ritualistic, recourse to the social context. On this view, the aim of science biography is to provide us with stories through which we can identify ourselves with other human beings who have chosen to spend their lives in scholarly or scientific work. Such stories can make us understand and change ourselves – scientists, historians of science and laymen alike. In that sense, biographies of scientists are useful as exemplars for what Kierkegaard considered to be the core of freedom in an ethical life: the continuous renewal of oneself.[148]

Hence, existential biographies of scientists may be edifying; they may provide us with opportunities for reorienting our familiar ways of thinking about our lives in unfamiliar terms, 'to take us out of our old selves by the

---

[144] For a discussion of authenticity in late modernity, see Taylor (1992).
[145] Klemm (1983).
[146] Cf. Nelson (1986), 465.
[147] Elliott (1990).
[148] Cf. Giddens' (1991) revival of Kierkegaard's vision in terms of 'life politics'.

power of strangeness, to aid us in becoming new beings'.[149] That is, by avoiding redescription of the scientist in the vocabularies of sociology or psychobiography, and by being sensitive instead to the vocabulary used by the scientist himself when trying to make sense of his life, biographers of scientists can, paradoxically, provide the exemplars we need for redescribing ourselves. Notwithstanding her programmatic flirtations with social biography, Keller's portrait of Barbara McClintock is primarily a story about what a marginalised and lonely woman's life in science was like. By retelling the struggles McClintock had to go through to succeed eventually in convincing other geneticists that the early molecular biologists' view of genetics was too simplistic,[150] Keller's portrait functions as a model for women scientists trying to cope with their own lives in science. Likewise Walter, by telling Bridgman's life story as a constant struggle with ethical choices in science, sensitises the reader to similar possible conflicts in his own life. In fact, both biographies, like Monk's Wittgenstein, provide examples of modern science hagiography – in the literal, not pejorative, sense of the word to be sure. Hagiography, as James McClendon points out in his discussion of theological biography, does not have to be understood as blind worshipping, but rather, 'at its best', as 'a mode of communal self-scrutiny'.[151] In that sense, I believe that one good existential biography of a scientist can better contribute to a remaking of the practices of science than a score of revealing social historical and sociological investigations of science.

Is this vision of science biography much different from that during the nineteenth century and the first half of the twentieth century, when the major purpose of the genre was to tell stories of the lives of great scientists,[152] or from Williams's wish, quoted above, to refocus on the great individual scientist? In one sense the answer is no – in fact, Robert Skidelsky has recently suggested that the biographer's main purpose indeed is 'to hold up lives as examples'.[153] He advocates biography as 'ancestor worship', as a genre that can recover the lessons older members of our community have made for us: 'The only way biography as an undertaking can recover its main function of good story-telling is to go back to . . . ancestor worship'. In another sense, however, a revival of this recurrent, but now widely rejected,

[149] Rorty (1980), 360.
[150] Keller (1983).
[151] McClendon (1990), x. I am grateful to Geoffrey Cantor for making me aware of McClendon's book.
[152] Theerman (1985).
[153] Skidelsky (1987).

theme of traditional science biography must certainly be very different. The hermeneutics of suspicion cannot, and should not, be undone. I am certainly not advocating a return to the uncritical hagiographic tradition, with its unqualified praise and glorification of the achievements of scientific 'giants'. Skidelsky uses the term 'ancestor worship' tongue-in-cheek: ancestor does not necessarily refer to a white, Anglo-Saxon, Protestant male, as Kenneth Manning's portrait of the black biologist Ernest Everett Just reminds us,[154] and worship does not refer to uncritical hagiography. The problem with traditional science biography was not that it provided personal models as such, but that these models were too bright and too unrealistic – stories of scientific heroes with whom it was difficult to empathise. What distinguishes existential biography from more traditional ancestor worship is the much greater range of lives to learn from, so that 'whereas in the past the exemplary principle worked in favour of tradition, today it works in favour of pluralism'.[155] Examplars do not have to be positive models. They can be negative, even Raskolnikovian, figures as well, models that teach us moral dilemmas, like John Heilbron's study of Max Planck as a lesson of 'heroic tragedy'.[156] Exemplars do not even have to have a great reputation; they can be ordinary members of the profession, like 'Anna'.[157] In fact, although a big contributor to science may provide a good example of the struggle between an ethical and an aesthetical life in science, lesser contributors (lesser egos) are probably more suitable to illustrate what it means to live an ethical life in science.[158]

Finally, even though the vision of science biography described here is motivated by a serious concern for commitment and edification, the style does not have to be boring. For a biography to be edifying does not exclude the possibility that it could be guided by stylistic consciousness. The time is ripe for science biographers to experiment with stylistic inventions such as collage, narrative discontinuity, multigenre narratives, unsuspected time-shifts, with stream of consciousness, symbolism, poetical reconstructions, and polyvocal texts, and so forth. In this respect, the art of science biography can also learn from film directors such as David Lynch (for example, *Wild at Heart*), Baz Luhrmann (*Strictly Ballroom*), or Wim Wenders (for example,

---

[154] Manning (1983).
[155] Skidelsky (1987), 1250.
[156] Heilbron (1986), viii.
[157] Goodfield (1981).
[158] Cf. Hull (1988), who claims that those scientists who are most selfish and egotistical are the greatest contributors to science, and conversely that the least productive scientists tend to behave the most admirably.

Wings of Desire) who are able to integrate an edifying message and a strong commitment to human values with stylistic inventiveness and an ironic distance to the plot and the characters.[159] Maybe we can even expect unconventional approaches such as that of Simon Schama, who relies heavily on imagination and impressionistic tales in bringing the past to life, to the extent that he challenges traditional notions of historical accuracy and reliability.[160] We can expect science biographies to become as adventurous and experimental as modernist novels — Russel McCormmach's fictional *Night Thoughts of a Physicist* being a pioneer example[161] — and particularly contemporary movies. In conclusion I see no reason why postmodern playfulness should not be able to coexist with the view of existential science biography advocated here.

Acknowledgements

An earlier version of this article was presented in the session 'The Scientist as Subject' organised by Geoffrey Cantor at the Second British–North American History of Science Meeting held in July 1992 in Toronto, Ontario. I am grateful to Inger Ravn for constant intellectual support, to Timothy P. Jackson, Michael Shortland, Skuli Sigurdsson and Richard A. Watson for criticising successive manuscript versions, to Ingemar Bohlin, Geoffrey Cantor, Kostas Gavroglu, Frederic L. Holmes, Johnny Kondrup, and Barbara G. Rosenkrantz for comments and suggestions, and to Craig Stillwell for linguistic corrections. Part of the research for this article was supported by a grant from the Swedish Research Council for the Humanities and Social Sciences.

Bibliography

Abir-Am, P. (1991) Noblesse oblige: lives of molecular biologists. *Isis*, 82, 326–43.
Anderson, D.D. (1981) Another biography? For God's sake, why? *The Georgia Review*, 35, 401–6.
Anon. (1985) Ich höre gerne zu. *Roche Magazin* (Basel: Hoffman-La Roche) 2, 2–17.
Barthes, R. (1977) *Roland Barthes by Roland Barthes*. New York: Farrar.
Bernstein, J. (1985) The merely personal. *American Scholar*, 54, 295–302.
Bloom, H. (1973) *The Anxiety of Influence*. New York: Oxford University Press.

---

[159] A small step in this direction is provided by Holmes (1991) who introduces a voice-over element in his Krebs biography.
[160] Schama (1991).
[161] McCormmach (1982).

Bourdieu, P. (1975) The specificity of the scientific field and the social conditions of the progress of reason. *Social Science Information*, 14, 19–47.

Bradbury, M. (1992) The bridgable gap. *The Times Literary Supplement*, January 17, 7–9.

Buber, M. (1958) *I and Thou*. New York: Scribner.

Cantor, G. (1991) *Michael Faraday: Sandemanian and Scientist: A Study of Science and Religion in the Nineteenth Century*. New York: St Martin's Press.

Cassidy, D.C. (1991) *Uncertainty: The Life and Science of Werner Heisenberg*. New York: W.H. Freeman.

Christie, J.R.R. and Orton, F. (1988) Writing on a text of the life. *Art History*, 11, 545–64.

Clifford, J. (1978) 'Hanging up looking glasses at odd corners': ethnobiographical prospects. In D. Aaron (ed.) *Studies in Biography*. Cambridge, MA: Harvard University Press, pp. 41–56.

Cohen, I.B. (1987) Alexandre Koyré in America: some personal reminiscences. *History and Technology*, 4, 55–70.

Cooper, D.E. (1990) *Existentialism: A Reconstruction*. Oxford: Blackwell.

Cooper, D.E. (1988) Life and narrative. *International Journal of Moral and Social Studies*, 3, 161–72.

Corbellini, G. (1990) *L'evoluzione del Pensiero Immunologico*. Torino: Bollati Boringhieri.

Cranshaw, R. and Lewis, A. (1989) Wilful ineptitude [review of Gowing *et al.*, *Cézanne: The Early Years*, 1988]. *Art History*, 12, 129–35.

Dash, J. (1973) *A Life of One's Own: Three Gifted Women and the Men They Married*. New York: Harper & Row.

Daston, L. (1995) The moral economy of science. *Osiris*, 10, 3–24.

Desmond, A. and Moore, J. (1991) *Darwin*. London: Michael Joseph.

Dilthey, W. (1989) *Introduction to the Human Sciences*. Princeton: Princeton University Press.

Dresden, M. (1988) *H.A. Kramers: Between Tradition and Revolution*. Berlin: Springer Verlag.

Dresden, M. (1990) A life in physics [Review of H.Kragh, *Dirac*, 1990]. *Science*, 249, 24 August, 937.

Eakin, P.J. (1985) *Fictions in Autobiography: Studies in the Art of Self-Invention*. Princeton: Princeton University Press.

Eakin, P.J. (1992) *Touching the World: Reference in Autobiography*. Princeton: Princeton University Press.

Eickelman, D.F. (1985) *Knowledge and Power in Morocco: The Education of a Twentieth-Century Notable*. Princeton: Princeton University Press.

Eiduson, B.T. (1962) *Scientists: Their Psychological World*. New York: Basic Books.

Eliot, T.S. (1960) Hamlet and his problems. In *The Sacred Wood*. London: Methuen.

Elliott, C.A. (1990) Collective lives of American scientists: an introductory essay and a bibliography. In E. Garber (ed.), *Beyond History of Science: Essays in Honor of Robert E. Schofield*. Bethlehem: Lehigh University Press, pp. 81–92.

Epstein, W.H. (1987) *Recognizing Biography*. Philadelphia: University of Pennsylvania Press.

Ferreira, M.J. (1989) Repetition, concreteness, and imagination. *Philosophy of Religion*, 25, 13–34.

Feuer, L.S. (1978) Teleological principles in science. *Inquiry*, 21, 377–407.

Forman, P. (1991) Independence, not transcendence, for the historian of science. *Isis*, 82, 71–86.

Frank, G. (1985) 'Becoming the other': empathy and biographical interpretation. *Biography*, 8, 189–210.

Gholson, B. Shadish; W.R. Jr, Neimeyer R.A. and Houts A.C. (eds) (1989) *Psychology of Science: Contributions to Metascience*. Cambridge: Cambridge University Press.

Giddens, A. (1991) *Modernity and Self-Identity; Self and Society in the Late Modern Age*. Stanford: Stanford University Press.

Golinski, J. (1990) The theory of practice and the practice of theory: sociological approaches in the history of science. *Isis*, 81, 492–505.

Goodfield, J. (1981) *An Imagined World: A Story of Scientific Discovery*. New York: Harper & Row.

Goodman, N. (1978) *Ways of World-Making*. Indianapolis: Hacket.

Gruber, H.E. (1981) *Darwin on Man: A Psychological Study of Scientific Creativity*. Chicago: University of Chicago Press.

Hammond, A.L. (1984) *A Passion to Know; 20 Profiles in Science*. New York: Scribner.

Hankins, T.L. (1979) In defence of biography: the use of biography in the history of science. *History of Science*, 17, 1–16.

Harré, R. (1986) *The Social Construction of Emotions*. Oxford: Basil Blackwell.

Heilbron, J.L. (1986) *The Dilemmas of an Upright Man*. Berkeley: University of California Press.

Holmes, F.L. (1991) *Hans Krebs: The Formation of a Scientific Life*. New York: Oxford University Press.

Holton, G. (1970) The roots of complementarity. *Dædalus*, 99, 1015–55.

Hull, D.L. (1988) *Science as a Process: An Evolutionary Account of the Social and Conceptual Development of Science*. Chicago: University of Chicago Press.

Jammer, M. (1966) *The Conceptual Development of Quantum Mechanics*. New York: McGraw-Hill.

Jerne, N.K. (1966) The natural selection theory of antibody formation; ten years later. In J. Cairns, G.S. Stent and J.D. Watson (eds.) *Phage and the Origins of Molecular Biology*. Cold Spring Harbor Laboratory of Quantitative Biology.

Jerne. N.K. (1971) The somatic generation of immune recognition. *European Journal of Immunology*, 1, 1–9.

Johnson, M. (1987) *The Body in the Mind: The Bodily Basis of Meaning, Imagination, and Reason*. Chicago: University of Chicago Press.

Keller, E.F. (1983) *A Feeling for the Organism: The Life and Work of Barbara McClintock*. New York: W.H. Freeman.

Kendall, P.M. (1986) 'Walking the boundaries'. In S.B. Oates (ed.) *Biography as High Adventure; Life-Writers Speak on Their Art*. Amherst: The University of Massachusetts Press, pp. 32–49.

Kierkegaard, S. (1946) *Either/Or*, 2 vols. Princeton: Princeton University Press.

Kierkegaard, S. (1992) *Concluding Unscientific Postscript to Philosophical Fragments*. Princeton: Princeton University Press.

Klemm, D.E. (1983) *The Hermeneutical Theory of Paul Ricoeur; A Constructive Analysis*. London: Associated University Press.

Kragh, H. (1989) *An Introduction to the Historiography of Science*. Cambridge: Cambridge University Press.

Kragh, H. (1990) *Dirac; A Scientific Biography*. Cambridge: Cambridge University Press.

Krauss, R.E. (1985) In the name of Picasso. In R.E. Krauss, *The Originality of the Avant-Garde and Other Modernist Myths*. Cambridge, MA: MIT Press, pp. 23–40.

Le Roy Ladurie, E. (1979) *The Territory of the Historian*. Hassocks, Sussex: Harvester Press.

Latour, B. (1987) *Science in Action: How to Follow Scientists and Engineers Through Society*. Milton Keynes, UK: Open University Press.

Lenoir, T. (1987) The Darwin industry. *Journal for the History of Biology*, 20, 115–30.

Lewontin, R.C. (1986) Review of D. Kevles, *In the Name of Eugenics*, 1985. *Isis*, 77, 314–17.

McClendon, J.W. Jr (1990) *Biography as Theology: How Life Stories can Remake Today's Theology*. Philadelphia: Trinity Press International.

McCormmach, R. (1982) *Night Thoughts of a Classical Physicist*. Cambridge, MA: Harvard University Press.

MacIntyre, A. (1982) *After Virtue*. London: Duckworth.

Manning, K.R. (1983) *Black Apollo of Science: The Life of Ernest Everett Just*. New York: Oxford University Press.

Manuel, F.E. (1968) *A Portrait of Isaac Newton*. Cambridge, MA: Belknap Press.

Miller, J. (1993) *The Passion of Michel Foucault*. New York: Simon & Schuster.

Monk, R. (1990) *Ludwig Wittgenstein: The Duty of Genius*. New York: The Free Press.

Morus, I.R. (1990) Industrious people: biography and nineteenth century physics. *Studies in the History and Philosophy of Science*, 21, 519–25.

Moulin, A.M. (1991) *Le dernier langage de la médicine: histoire de l'immunologie de Pasteur au Sida*. Paris: Presses Universitaires de France.

Nadel, I.B. (1984) *Biography: Fiction, Fact and Form*. New York: St Martin's Press.

Nelson, W. (1986) Sabato's *El Tunel* and the Existential Novel. *Modern Fiction Studies*, 32, 459–67.

Nye, D. (1983) *The Invented Self: An Autobiography from Documents of Thomas A. Edison*. Odense: Odense University Press.

Pernis, B. and Augustin A.A. (1982) Review of C. Steinberg and I. Lefkovits, *The Immune System: A Festschrift in Honor of Niels Kaj Jerne. European Journal of Immunology*, 12, 1–3.

Polanyi, M. (1958) *Personal Knowledge: Towards a Post-Critical Philosophy*. Chicago: University of Chicago Press.

Polanyi, M. (1959) *The Study of Man*. Chicago: University of Chicago Press.

Polanyi, M. (1966) *The Tacit Dimension*. Garden City: Doubleday.

Popper, K. (1974) Autobiography of Karl Popper. In P.A. Schilpp (ed.) *The Philosophy of Karl Popper*. (La Salle, IL.: Open Court.

Provine, W.B. (1986) *Sewall Wright and Evolutionary Biology*. Chicago: University of Chicago Press.

Ricoeur, P. (1970) *Freud and Philosophy*. New Haven: Yale University Press.

Ringer, F. (1990) The intellectual field, intellectual history, and the sociology of knowledge. *Theory and Society*, 19, 269–94.

Rorty, R. (1980) *Philosophy and the Mirror of Nature*. Princeton: Princeton University Press.

Rorty, R. (1989) *Contingency, Irony and Solidarity*. Cambridge: Cambridge University Press.

Rosenberg, C. (1988) Woods or trees? Ideas and actors in the history of science. *Isis*, 79, 565–70.

Runyan, W.M. (1984) *Life Histories and Psychobiography: Explorations in Theory and Method*. New York: Oxford University Press.

Runyan, W.M. (ed.) (1988) *Psychology and Historical Interpretation*. New York: Oxford University Press.

Schama, S. (1991) *Dead Certainties*. New York: Knopf.

Scriven, M. (1984) *Sartre's Existential Biographies*. London: Macmillan.

Shapin, S. (1992) Discipline and bounding: the history and sociology of science as seen through the externalism–internalism debate. *History of Science*, 30, 333–69.

Shapiro, M.J. (1986) Reading biography. *Philosophy of the Social Sciences*, 16, 331–65.

Sheets-Pyenson, S. (1990) New directions for scientific biography: the case of Sir William Dawson. *History of Science*, 28, 399–410.

Skidelsky, R. (1987) Exemplary lives. *The Times Literary Supplement*, November 13, 1250.

Skinner, B.F. (1984) *Particulars of My Life, The Shaping of a Behaviorist*, and *A Matter of Consequence*. New York: New York University Press.

Smith, C. and Wise N. (1989) *Energy and Empire: A Biographical Study of Lord Kelvin*. New York: Cambridge University Press.

Smocovitis, V.B. (1991) The politics of *Writing Biology*. *Journal of the History of Biology*, 24, 521–7.

Sokal, M.M. (1990) Life-span developmental psychology and the history of science. In Elizabeth Garber (ed.), *Beyond History of Science: Essays in Honor of Robert E. Schofield*. (Bethlehem, PA: Lehigh University Press.

Söderqvist, T. (1991) Biography or ethnobiography or both? Embodied reflexivity and the deconstruction of knowledge-power. In F. Steier (ed.), *Research and Reflexivity*. London: Sage, pp. 143–62.

Tauber, A.I. and Chernyak, L. (1991) *Metchnikoff and the Origins of Immunology: From Metaphor to Theory*. New York: Oxford University Press.

Tauber, A.I. (1994) The Immune Self: Theory or Metaphor. Cambridge: Cambridge University Press.

Taylor, C. (1992) *The Ethics of Authenticity*. Cambridge, MA: Harvard University Press.

Theerman, P. (1985) Unaccustomed role: the scientist as historical biographer – two nineteenth-century portrayals of Newton. *Biography*, 8, 145–62.

Unger, R.M. (1984) *Passion: An Essay on Personality*. New York: The Free Press.

Wallulis, J. (1990) *The Hermeneutics of Life History; Personal Achievement and History in Gadamer, Habermas, and Erikson*. Evanston: Northwestern University Press, 1990.

Walter, M.L. (1990) *Science and Cultural Crisis; An Intellectual Biography of Percy Williams Bridgman (1882–1961)*. Stanford: Stanford University Press.

Watson, J.D. (1968) *The Double Helix: A Personal Account of the Discovery of the Structure of DNA*. London: Weidenfeld and Nicolson.

Westfall, R.S. (1980) *Never at Rest: A Biography of Isaac Newton*. Cambridge: Cambridge University Press.

Williams, L.P. (1991) The life of science and scientific lives. *Physics*, 28, 199–213.

Young, R.M. (1988) Biography: the basic discipline for human science. *Free Associations*, 11, 108–30.
Young-Bruehl, E. (1982) *Hannah Arendt: For Love of the World*. New Haven: Yale University Press.

2

Life-paths: autobiography, science and the French Revolution

DORINDA OUTRAM

Introduction

François-René de Chateaubriand, the great French romantic, spoke for an entire generation when in his autobiography he described the French Revolution as a 'river of blood which separates for ever the old world in which you were born from the new world on whose frontiers you will die'.[1] This sense of the French Revolution as a dark and threatening chasm in time, so different from the joyful *novus ordo saeculorum* of the triumphant New World revolution of 1776, was undoubtedly one of the major reasons for the outpouring of autobiography which is characteristic of that era.[2] This was so not only because autobiography seemed the ideal vehicle, in many cases, to vindicate the actions of the troubled revolutionary era, but also because autobiography became a way of asserting the continuities of existence and identity across the rupture between the pre- and post-revolutionary worlds. Because of this, it is difficult, and misleading, to see the autobiographies of this time as 'private' documents: rather, they are involved with the reshaping of the meaning of the revolutionary era, precisely because they are the medium through which many in France worked through the specific nature of their investment in, and exploration of, the 'new world' of post-revolutionary time.

Possibly nowhere did autobiography fill this need more urgently than in the case of the scientific élite. As the old élites emigrated, were executed, or were forced out of employment, the élite of science became one of the few

---

[1] Chateaubriand (1958), 165.
[2] Dekker (1969), 61–72; 63–4; Tulard (1971) lists 794 items of printed autobiography. The collection omits MS autobiography, and has little interest in the scientific community. The growth of introspective private writing of all kinds is described in Bryant (1978).

sources of the technical expertise so greatly needed by successive French regimes forced to mobilise the resources of France in the warfare which began in 1792 and was not to end until 1815. It was this which allowed the hitherto unprecedented rise of scientific men to positions of real political and administrative power.[3] This capture of power outside the narrow area of technical expertise was aided by the fact that the public culture of the Revolution legitimated its break with the past by appeals to values which science was uniquely fitted to endorse: by appeals to 'Nature', and to 'Reason', as distinct from the key cultural values of the former monarchy, of historicity, heterogeneity and artificiality. For a revolution trying to find ways in which it could do that which the Jacobin regicide Louis-Antoine St Just defined as impossible for monarchy and 'reign innocently', science was the ideal cultural legitimation, and its practitioners therefore legitimate bearers of public authority.[4]

Science, and its individual practitioners, thus underwent a dramatic change in cultural field and force. At the same time many individuals within the scientific community experienced extreme reversals of personal fortune. It is thus hardly surprising that autobiographical writing should flourish among the French scientific élite of the Revolutionary period, as a means of coming to terms with the unbridgeable spaces opening up between past and present lives and roles. This outburst of autobiographical writing was also stimulated by a specific institutional pressure. Georges Cuvier and Jean-Baptiste Delambre, the Permanent Secretaries of the First Class of the Institut, made it a practice to request autobiographical material from those whose funeral orations or *éloges*, they might later have to write.[5] However, the fact that many, though not all, autobiographies were produced on request from those whose task it was to manufacture the 'official history' of science, does not seem to have made such autobiographies bland or impersonal; far from it. In fact, one of the most interesting features of the autobiographies is their very difference from the official lives of their subjects. This is a topic to which I will return.

But the fact that autobiography became in any way a resource for *éloges*, in itself signals the increasing incidence and acceptability of autobiographical writing among the scientific élite. For the efflorescence of autobiography

---

[3] Outram (1984 and 1980).

[4] For artifice in the public life of the old Regime, see Sennett (1974), 1–99; Louis-Antoine St Just, speech of 13 November 1792, quoted in Walzer (1974), 124, from *Archives Parlementaires*, première série, 53 (1792), 391.

[5] Autobiographies produced in this context are listed in Outram (1978).

amongst this group was not unchallenged and did not have deep historical roots. More traditional thinkers still insisted that for men of learning, autobiography was not only unnecessary, but was even a logical contradiction to the ideal of the identity of life and work. The absolute identity of life and work, and hence the inutility of autobiography, for all *savants*, was still being maintained as late as 1807, when Pierre-Charles Levesque maintained of the classical scholar Apollonius of Tyanus that 'it seems that following the precept of an ancient philosopher he wanted to hide his life; but he completely revealed it when he published his work'. This identity could only be damaged by the process of introspection so necessary to autobiography, with its implication that the path from past to present, from the life to the work, was likely to be broken, uneven, and full of the unforeseen.[6] The social and intellectual conditions of autobiography in the French scientific élite could not therefore rely on a settled social response. The age itself challenged the idea of the centrality of the individual; the professional ethos was unsure of its reception of autobiography.

The French Revolution had produced perceptions of the fundamental category of autobiography, time itself, which were new and disturbing. The idea of the Revolution as a rupture in time, was accompanied by an unparalleled sensation of time as having speeded up.[7] Perhaps we are seeing here the first symptoms of what modern theorists of the information revolution have described as the collapse of the categories of real time. In this situation, writing autobiography, which depends on the unfolding of lived time, represented severe challenges. This paper will examine several examples of autobiography produced by members of the French scientific élite to show how autobiographical writing responded to these varied pressures. The examples discussed in this paper are drawn from only a small fraction of extant autobiographical writings. These are listed these at the end of this chapter; however many more undoubtedly await discovery in family and public archives. Most of these autobiographies were produced by members of the French scientific élite, members of the Academy of Sciences, or the National Institute. Far from being private documents, these autobiographies were often produced

---

[6] Even the 'autobiography' of G.B. Vico maintains an identity between 'life' and 'work', in spite of the fact that it was also Vico who began to generate and collect scientific autobiography (Fisch and Bergin (1944)); Levesque (1807), xviii: 'il semble que, soumis au précepte d'un ancien philosophe, il ait voulu cacher sa vie; mais il l'a dévoilée toute entière en publiant ses ouvrages'.

[7] E.g. Mme Roland exclaimed, 'Here we live ten years in twenty-four hours': Perroud (1900–1901); Ramond remarked in the National Assembly 'the last six months of 1789 were worth an ordinary year' (2 January 1792) quoted in Aulard (1885), I, 89.

to respond to public or institutional demand. This is an important point to make because autobiography can often be used as the primary source for biography. It is thus important to realise that autobiographies are not raw data but reflect the traces of the many pressures on autobiographical writing; as a genre, they are a precious resource but also tell us much about the problems of the scientific enterprise in revolutionary and post-revolutionary France.

Journeying, knowledge and self-knowledge

To confront the central metaphor of my title: the image of life as a path to be followed is a very ancient one, which was transformed in the Christian West into the image of man as a lifelong pilgrim. Debate raged by the end of the Middle Ages over the proper conduct of the pilgrim, and involved issues of particular importance for the sciences. At the heart of the debate on pilgrimage was the problem of curiosity.[8] Pilgrims both actual and metaphorical were repeatedly cautioned not to let their eyes stray from their paths, if they were to arrive at their goal. This is an image still active in Protestant texts of the seventeenth century such as John Bunyan's *Pilgrim's Progress*. By the eighteenth century, this debate had been secularised and inverted. Journeying, and the growth of knowledge in and about life, were repeatedly connected. Jean-Jacques Rousseau's *Emile* (1762), for example, produced a strong defence of straying from the beaten path as the best way to acquire knowledge:

> The only way I can imagine travelling more pleasant than horseback) is to go on foot. . . . You can see all the countryside, you can turn left or right, you can have a proper look at anything which interests you, and you can stop wherever there is a view. If I notice a river I walk along its banks; if a I see a deep wood, I enter its shade; a cavern, I can enter it; a quarry, I can look at the rocks. To travel on foot is to travel like Thales, Plato, and Pythagorus. I really can't understand how an enquiring person could decide to travel in any other way.[9]

---

[8] Ladner (1967); Zacher (1976).

[9] Rousseau (1964), 522–3: 'Je ne conçois qu'une manière de voyager plus agréable que d'aller à cheval; c'est d'aller à pied. On observe tout le pays; on se detourne à droit, à gauche; on examine tout ce qui nous flatte; on s'arrête à tous les points de vue. Aperçois-je une rivière, je la cotoie; un bois touffu, je vais sous son ombre; une grotte, je la visite; une carrière, j'examine les mineraux . . .

While this passage makes very clear the eighteenth-century links between straying from a set path and observational curiosity, later, the very title of Rousseau's *Reveries of a Solitary Walker*[10] (1782) was to make the connection between covering the ground and the work of introspection even more complete. It was a metaphor which also allowed the linkage of the life and work to go on being made at another level not by cutting out the life, but by seeing it as a web of movement, curiosity and introspection, which came together in a scientific vocation.

It was in this way that movement and autobiography became inextricably linked, and made their way into the autobiographies of men of science. Sometimes, accounts of journeying are also ways of dealing with the fault lines which described an entire life. For Cuvier, for example, the inner movement from childhood to adolescence was also a movement from one language to another, all encapsulated in an actual journey from his birthplace in provincial Montbéliard, to school in cosmopolitan Stuttgart. Such was the impact of this linked inner and outer journey, that even forty years later, Cuvier was unable to recall it without experiencing

> a sort of terror at this journey, which I made in a little carriage
> squashed between the duke's chamberlain and his secretary, who
> I was really in the way of, because there was barely room enough
> for them as it was. For the whole journey they only conversed in
> German, which I didn't understand a word of, and hardly spoke
> a single word of encouragement or consolation to me.[11]

In the case of Bernard Germaine Lacépède, journeys too are paths into different self-representations. As his autobiography details his journeys from Paris to Agen, from Agen to Paris, from Versailles to Germany, it also details his travelling between the different roles which one after the other filled his life before his vocation as a man of science became firm. Local notable in Agen and erudite, in Paris, musician, in Germany, soldier and diplomat, Lacépède's vocation only became clear when he reached the end of a final voyage, from Paris to his wife's property at Montlhéry:

Voyager à pied, c'est voyager comme Thalès, Platon et Pythagore. J'ai peine à comprendre comment un philosophe peut se resoudre à voyager autrement'. A further panygyric to curiosity, linking it to movement, is on p. 185.

[10] Written between 1776 and 1778, and published in 1782. The same connection between walking, 'pilgrimage' and self-knowledge persisted into the following century: Thoreau (1862).

[11] The autobiography of Cuvier is MS Flourens 2598 of the Library of the Institut de France: 'une sorte d'effroi, à ce voyage que je fis dans une petite voiture, entre le chambellan et le secrétaire du Duc, que je gênais beaucoup, parcequ'il y avoit à peine de la place pour eux, et pendant toute la

> It was in the beautiful countryside which surrounds this village, in
> the charming meadows, in the little woods near the old country
> house, that I thought through the plan for my natural history, and
> here too that I wrote most of its first volume.[12]

The metaphor and reality of travel as being and becoming, of life as a curious exploration of many paths, is extended in the autobiography of Ramond de Charbonnières.[13] A tumultuous early life, always following 'un autre horizon',[14] was almost brought to a close in the prisons of the Terror, where Ramond compared himself to Robinson Crusoe, the voyager and spiritual pilgrim, who in losing all upon his desert island, finds the spiritual insight he had previously disdained.[15]

In travelling towards their making and remaking of self, autobiography writers of the eighteenth century were not alone. They travelled with a whole range of different reference figures on which to base the struggle for identity. The eighteenth century was an age of mimesis – an age when modelling the self on some figure of 'representative virtue' was a commonplace psychological exercise.[16] For no group was it an exercise more important than for the scientific community, the one most in need of models to negotiate the transition between old and new worlds. This means also that in considering eighteenth-century autobiography, we are often not merely considering the life of the ostensible subject, but also the relationship between that life and the role models which he or she adopts.

But the search for role model also transmitted conflict into the autobiographical form. Role models could be the stern figures of classical antiquity, or the newer heroes of science and virtue such as Newton or Benjamin Franklin. They could also be taken from the emotional narratives which were eighteenth-century best-sellers, novels like Richardson's *Clarissa* (1747–8) or Rousseau's *Julie ou la Nouvelle Héloïse* (1761). These were models whose actions put them fairly at odds with the restraint and disinterest expected by the ideal man of science; they also encapsulated, in their exaltation of emotion, a whole response to the natural world which would be characteristic

---

route ne se parlèrent qu'en allemand, dont je n'entendais pas un mot, et m'addressèrent à peine deux paroles d'encouragement et de consolation'.

[12] Hahn (1975), 61, 64: 'C'était dans les belles campagnes qui environnent ce village, dans ses charmantes prairies, dans les bosquets de l'ancien château, que j'avais médité le plan de mon *Histoire naturelle*, et composé une grande partie du premier volume de cette Histoire'.

[13] Dehérain (1905).

[14] Ibid., 124.

[15] Ibid., 128. 'ah! le beau livre que je vous ferais, nouveau Robinson que j'ai été . . .'.

[16] Tatin (1985).

of Romanticism. This is a conflict epitomised by the difference between the Rousseau of the *Confessions*, of *Julie* and of the *Reveries*, and the Rousseau of *Emile* and the *Social Contract*.[17]

These were conflicts of which autobiography writers were well aware. Ramond, for example, consciously modelled his autobiography in novel form to avoid both the very question of the identity of life and work, as well as to avoid the political commitments implicit in the sterner genre of the *Mémoire* or 'History of my own times'. He wrote to his friend St Amans, who had urged him to produce an autobiography:

> you want me to stop writing about Nature, and become part of the throng of those who are constantly retailing the slightest doings of our contemporaries. You need a 'life and times', you aren't put off by all this; but my life and times? Certainly they would amuse a lot of people, even though it would be my puny self who would be playing the hero for, as well as as the great and small events which I would be describing as so many others have done, the ups and downs of my own personal fortunes would read like a novel and have all the pulling power of the genre.[18]

The autobiography which resulted is, in fact, written as a picaresque novel. To take another example, Lacépède wrote some passages of his autobiography in a way which recalls nothing more than the reactions to scenery of the characters in *Julie*:

> Seated on the ruins which surround the high tower at Montlhéry, looking out over an immense prospect, seeing the distant outline of the monuments of the capital; or lying on the flowered grass, in the shade of whispering poplars on the shores of the great lake at Marcoussis, or strolling under the green vault of the vast and lonely forest on the mountain tops around the lake, I love to meditate on the wonderful effects of the power of Nature . . . Given over to these sublime thoughts, dragged away by great conceptions,

---

[17] Outram (1989).

[18] Dehérain (1905), 122: 'vous voudriez que j'abandonasse cette partie des ouvres de la nature qui va sans dire, pour me mêler dans la foule de ceux qui bavardent dans tous les sens sur les faits et gestes des hommes de leur tems. Il vous faut des mémoires: vous n'êtes pas dégouté: mes mémoires! certes ils amuseraient beaucoup de monde, quand même mon chétif individu y serait sur le premier plan, car au milieu des grands et petits evenemens dont j'aurais à essayer l'esquisse après tant d'autres, les vicissitudes de ma propre fortune constitueraient à elles seules une manière de roman qui aurait tout l'intérêt du genre'; Trahard (1936), 31–6.

seduced by these magical scenes, I forgot the world, and saw
nothing but the universe.[19]

Here, Lacépède portrays his relation with Nature as very different from that
of the self-sufficient, austere man of science beloved of the Stoic tradition.
His attitude is defined by emotion, not reason, by a blurring of the boundaries between himself and Nature due to an emotional connection with
landscape. In this part of his autobiography, Lacépède becomes the man of
sentiment so beloved of the eighteenth-century novel. The smooth self-
possession of the Stoic ideal vied for place with Rousseau's ideal of autobiography as a total revelation, and life as the fashioning of emotion.[20] Lacépède in fact uses autobiography as a story of the clarification of commitment,
to science, a commitment which seems, at many points in the autobiography, to be nearly overwhelmed by the pressure of unresolved emotion.

Autobiography: vehicle for commitment

Clearly, autobiography was of great utility in clarifying the commitments of
the writer. And it is not surprising that this psychic function should be mir-
rored by a focus on episodes of commitment within the autobiographies
themselves. To some extent, such episodes can be recounted as moments of
absorption, like the one just quoted from the autobiography of Lacépède; or
the one contained in the otherwise very different autobiography of Georges
Cuvier, who relates how Buffon's Natural History completely absorbed him
as a child: 'All the pleasure I had as a child was in copying its pictures and
colouring them according to the captions'.[21] Such moments of intense focus,
with their links to later self-dedication, have obvious religious echoes.[22] And

[19] Hahn (1975), 69: 'Assis sur les ruines qui environnent la haute tour de Montlhéry, dominant sur
un pays immense, découvrant de loin le faite des superbes monuments de la capitale, ou couché sur
un gazon fleuri à l'ombre de peupliers inspirateurs et sur les bords du grand étang de Marcoussis, ou
me promenant sous les vôutes de verdure formés par les vastes et solitaires forêts qui courronnaient
les montagnes autour de cet étang, j'aimais à méditer sur les admirables effets de la puissance de la
nature . . . Livré à des conceptions élèves, entraînée par ces grandes pensées, séduit par ces tab-
leaux magiques, j'oubliai le monde, je ne voyais plus que l'univers.'

[20] For the influence of Rousseau's idea of autobiography as complete self-revelation, see Outram, The
Body (note 17), 149–51.

[21] Discussed in Outram (1984), 19, and quoted from MS Flourens 2598 of the Institut de France:
'Tout mon plaisir d'enfant était d'en copier les figures, et de les enluminer d'après les descriptions'.
For the importance of the portrayal of such moments of absolute concentration of the forces of the
personality, see Fried (1980).

[22] For modern equivalent experiences, see Terres (1961); Cobb (1977); Austin (1931), 24–5: 'It was
a summer morning, and the child I was had walked down through the orchard, and come out on
the brow of a sloping hill where there were grass and a wind blowing, and one tall tree reaching

it is worth mentioning at this point how closely autobiography had previously been linked with the history of individual spirituality. Indeed, in
many national traditions, autobiography *was* the recounting of spiritual pilgrimage and conversion.[23] In spite of the secularisation of many features of
autobiography during the eighteenth century, and its mixing, as we have
seen, with other forms of narrative of individual lives it remained marked by
the earlier history of the genre. Moments of epiphany, of absorption in
Nature, in scientific autobiography have the same role as conversion
moments in spiritual autobiography: they resolve the antagonism of the self
and the world. What brings optimism to Lacépède's account is the way in
which his autobiography is structured as a journey *towards* such an epiphany: namely, his absorption in Nature. What gives Cuvier's its pessimism
is the movement of his account away from 'absorption' in infancy, into a
relentless confrontation between himself and the 'world', power and patronage, conceived as something whose control and manipulation must be so
endlessly recreated that such contact with Nature, such blurring of the lines
of the self, can never recur. The precision with which Cuvier tells us the
details of this conflict is precisely what makes Cuvier's autobiography such
an important source in the history of scientific politics and patronage.

Nor is this the only way in which even the post-Rousseau autobiography
retains traces of the religious associations of the genre. Linking science,
religion and autobiography was the notion of vocation. Often, the finding of
a vocation is linked with the rejection or loss of a biological father, and the
finding of new, 'social' father, the patron, who guides the novice man of
science and acts as his role model. Just as in early Christianity, the finding of
a religious vocation, the rejection of the biological family, and the adoption
of a saint as patron, were often inextricably mixed events.[24] This theme is
particularly marked in Lacépède's autobiography, which describes many
moments of 'adoption' by individual patrons. Lacépède begins with his childhood memories of the bishop of Agen, who treated him 'as if I had been his
child',[25] as a young man Buffon regarded him 'comme son fils', and after the

---

into infinite immensities of blueness. Quite suddenly after a moment of quietness there, earth and
sky and windblown grass, and the child in the midst of them came alive together with a pulsing
light of consciousness. There was a wild foxglove at the child's feet, and a bee dozing about it, and
to this day (1931) I can recall the swift inclusive awareness of each for the whole – I in them and
they in me and all of us enclosed in a warm lucent bubble of livingness'.

[23] Dekker (1969); Niggl (1977); Roodenburg (1985).
[24] Theis (1976); Silverman (1974).
[25] Hahn (1975), 56.

devastating event of his own father's death in 1783,[26] he immediately formed a close relationship with the Gauthier family, who virtually adopted him.[27] Later, he was to marry the widowed Mme Gauthier, and it was, as we have seen, on her estate, that his vocation as a natural historian was solidified: absorption in Nature, and reintegration into a second, 'social' family, came together. A similar process was undergone by the young chemist Eleuthère Dupont de Nemours.[28] So common was it, that Cuvier's autobiography is at great pains to explain why it was that, unlike his peers, he was unable to record the same moments of adoption and confirmation by a patron. Clearly, such moments of rupture with natural family, and readoption by 'social' family or patron taking the role of father fitted easily with Stoic ideals of the essential disinterest of the man of science, and with a vision of his role as outside social production.

Autobiography under the conditions of revolution

But these were roles maintained against great odds. The period of the Revolution saw a collapse, for all sections of the French élite, of the patronage networks stemming from members of the former privileged orders. With these networks in disarray, the search for patronage and recognition became ever more uncertain, and this uncertainty was compounded by the fact that the role of the man of science in public life was in rapid transition. It was also a victim of the insecurity of Revolutionary politics, and the collapse of all accepted forms of status and social recognition. For those who fell out of political favour, the problem of maintaining a consistent face to the world thus became almost insuperable. Because of this, autobiographies from this period often detail dramatic shifts in social status, shifts which are perceived as changes in dramatic role. As Ramond comments, 'I could see the storm clouds coming nearer. I hoped to be able to be just a spectator; but events quickly made me an actor in the drama'.[29] For many in the scientific community, as outside it, survival itself often depended on the rapid conversion

---

[26] Ibid., 62: 'La douleur que j'eprouvai fut affreuse . . . et ma vive et bien juste affliction me donna une maladie nerveuse qui ne me permit de quitter Agen pour revenir à Paris, que vers le milieu de l'été 1784'.

[27] Ibid., 63: 'Ils me permirent de les regarder comme un frère et comme une soeur, et après les pertes irréperables que j'avois faites, combien ne dus-je pas me féliciter de retrouver en eux une seconde famille!'

[28] Dupont de Nemours (1906): 'Je n'étais qu'un enfant lorsqu'il me tendait les bras. C'est lui qui m'a fait un homme'. Cuvier's autobiography is virtually unique in its absence of such moments.

[29] Dehérain (1905), 126: 'je voyais approcher l'orage. Le rôle de spectateur était celui où j'espérais pouvoir me renfermer: la force des choses me mit bientôt au nombre des acteurs'.

from the roles of the *Ancien Régime* into those less conspicuous, or more politically correct.[30] Very often, such role changes were associated with a fall in social status. The naturalist Louis-Augustin Bosc, for example, too closely associated with the Girondins in 1793, recounts in his autobiography how he fled Paris, and adopted the clothes and life-style of a working man.[31]

Life under the revolution placed a survival value on mimesis. But these episodes of role-shift also interrupt the flow of many autobiographies from past to present, doubt to resolution. For some, the interruption in their role as scientific men, and accompanying status, was too great to bear.[32] For others, such as Ramond, science emerged as the lifeline which held identity together.[33] But the enforced roles of the Revolution inject an incongruous note into the strands which we have hitherto identified as part of the heritage of scientific biography. The idea of life as a journey, the idea of vocation as achieved in moments of 'epiphany', of supreme absorption, and as realised by the movement between natural family and social family, were all perturbed by the incongruous roles forced upon many men of science by the Revolution. Under such roles, the speaking subject of the autobiography becomes literally masked, the path of life diverted. The chasm in time which was for many the Revolutionary experience, also becomes a chasm in autobiographical narrative.

In registering this disruption, the autobiographical form was also marking another major change in life-paths. Before the revolution, the possession by a single individual of multiple social roles was not viewed as exceptional. This is precisely why many autobiographies detail a twisted and divergent path towards their subject's vocation. Nor does it surprise us that men like the chemist Antoine-Laurent Lavoisier should hold multiple social roles, roles which would now be seen as grossly incompatible should any individual try to combine them in the twentieth century. Landowner, financier, local administrator and research chemist, were combined in Lavoisier's person, and his case is far from unique.[34] The teasing out and separation of roles from each other was to be a characteristic of the nineteenth century, one which was to help to make autobiography a different exercise, partly

---

[30] Outram (1983).

[31] Bosc's autobiography is cited in an edited version in Perroud (1905): 'je m'habillai à la sans-culotte, travaillai à la terre, au bois . . . fis moi-même ma cuisine'.

[32] See autobiographical writings of the astronomer Cassini, discussed in Outram (1983), 263–44, using autobiographical material quoted in Devic (1851), 290–91.

[33] Outram (1983), 256.

[34] Little attention has been paid to this problem: see however, Vergnaud (1945–1955); Outram (1984), 93–117.

by changing the social conditions in which autobiography was written and received.

A final point to note about these autobiographies is that of their gendered Nature. All extant scientific autobiographies of this period known to me are written by men.[35] They recount experiences which were only available to men in this period, experiences such as the search for membership of élite scientific institutions, or the holding of public office. But in another way, even the autobiographies of men of science manifest much of the confusion and contradiction entering into pre-revolutionary experiences and discourses of gender. Writings such as Lacépède's, for example, are suffused with the emotionality of the novel, which, as literary critics have noted, increasingly made the literary field the producer of common emotional patterns and role models between the genders.[36] On the other hand, many other autobiographies, still revolve around the very masculine stereotype of the Stoic man of science. This is key to Ramond's self-portrayal at the end of his captivity, for example, as a 'new Robinson', denuded of the world's goods, and with the wish, after the passions of the Revolution, to focus only on the objectivity of Nature.[37] The Revolution itself was to see a dramatic hardening of the division between the genders, a hardening whose consequences for future politics, and for the way in which science was professionalised, have only recently come to be seriously appreciated.

It is also the case that these male autobiographies tend to produce a different account of the 'turning-points' of life. In many autobiographies the stage of life that makes the transformation from child to adult, is marked, as we have seen, by a moment of 'adoption' by an older, usually male patron, as well as moments of supreme absorption in Nature itself. The definition of vocation in other words, in most male autobiographies, is defined as finding a new 'father', becoming the 'child' again. Little stress is placed in many of the autobiographies on marriage or paternity, usually mentioned only in passing, if at all. In this respect, the autobiographies are very similar to the official contemporary biographies or *éloges*. This is a very different world from the life cycle events such as puberty, marriage and maternity, which we find dominant in contemporary female autobiography, such as that produced

---

[35] Women, however, often played a significant role in the editing of manuscript autobiographical material: the serious excisions in Cuvier's autobiography perpetrated by Mme Cuvier after his death in 1832, are a case in point. The influence of women on the preparation of documentary evidence was all the greater in the period of the Revolution due to the high number of male deaths.

[36] Eagleton (1982); Outram (1989), 138–41. For contrary argument see Nussbaum (1989).

[37] Dehérain (1905), 128.

by Mme Roland.[38] These are differences which reflect the very different social insertion and self-images of men and women at this time. Research into actual life situations amongst the scientific élites of this period reveals the very ambiguous role of marriage for the man of science, and the heterogeneous character of the marriage partners chosen.[39] Marriage in other words, was a social practice whose integration with the self-image of the man of science, was very problematic. Very often, this exclusion of the marriage relationship, or other love relationships, from both biography and autobiography, results in what would now be seen as bizarre displacements of personal emotion into the fabric of scientific writing. This means that it is difficult both to talk of 'scientific' writing in the same way that we would do now, as descriptions of investigations into Nature devoid of personal emotional content; and, that it is also wrong to draw rigid lines between autobiography and other forms of writing. Just as scientific professional roles were only starting to disentangle themselves from the multiple roles taken for granted by élites, so it was that autobiography itself was yet to separate out from other forms of scientific prose. Lacépède, for example, remarks matter-of-factly in his autobiography about his grief at the death of his wife in 1801: 'There is an expression of my grief in the fifth volume of my *Natural History of Fish*'.[40] It is clear from these examples that the impermeable divisions between the work and the life, Nature and its observer, objectivity and subjectivity, the product of 'male science' according to much feminist writing, were not yet in place in the writings of these male scientists.[41]

There is a real contradiction here between this play of violent personal emotion into scientific writing, and the ideal of the man of science as a calm and disinterested figure, moved only by the encounter with Nature.[42] This contradiction is a genuine one and can be traced back to the porosity of the autobiographical form itself, which was here only reflecting the cacophony of different models of identity with which Enlightenment men and women were surrounded, and tried to internalise. In turn, this reflected increasing problems in this period in defining both public and private roles for individuals. France at the end of the eighteenth century saw increasing competition

[38] Perroud (1905); Geiger (1986).
[39] Outram (1987), 19–30.
[40] Hahn (1975), 74; Lacépède adds to his autobiography an extract from his *Histoire de l'Europe* (1816) in which he recalls his grief on the death of his daughter-in-law (ibid., 83). Among many other possible examples could be cited Dolomieu (1793), 41, which contains his eyewitness account of the murder of his friend the Duc de la Rochefoucauld.
[41] For the social consequences of the absence of these divisions see Outram (1987).
[42] Sonntag (1974); Outram (1978).

to define the public realm, and thereby define who the actors in that public realm were to be:[43] who, in other words, were to replace the monarch as sole public actor. Increasingly, the definition of sovereignty, of public identity, became conflated with that of self-possession, self-definition. This may also be one of the reasons why most scientific autobiographies of this period are written by men who were far from the end of their life-span. These autobiographies were not primarily written to display the distillation of a lifetime search for achievement and self-knowledge; they were typically written in mid-life, to *create* public personalities and manifest self-possession. They were documents written to induce change, not to record passively that which had already occurred. In this way, they contrast strikingly with autobiographies by English men of science such as Humphry Davy, almost all of which were produced later in life.

But in saying all this we are making something more than the banal point that scientific autobiography, like any other form of art, reflects the strains of the age which produced it. We are also pointing to one of the very reasons for the resurgence of the scientific community in social, cultural and political authority in the 'new world' which emerged after 1789. To have resolved all these complex and contradictory elements into the search for a human identity which manifested the contradictory elements of the vocation of science, was to have absorbed too, a responsiveness to change, a capacity to intervene in the chaotic public culture of post-revolutionary France. To be able to offer its practitioners to the public both as practitioners of Stoic calm, and as mobilising an emotional relationship with Nature, was to offer a resource unique in the public culture of the age. They could do this because these autobiographies were still structured around the turning points of vocational choice, rather than around moments of discovery in science which were to become important in nineteenth-century biographies and autobiographies. Whereas eighteenth-century autobiographies are powerful vehicles which talk about mobilising a *life*, the nineteenth-century autobiographies focus on a single 'Eureka' *moment* of discovery, and fail to engage with their surrounding culture as a whole.

Bibliography

Abir-Am, P. and D. Outram (eds) (1987) *Uneasy Careers and Intimate Lives: Women in Science 1789–1979*. New Brunswick: Rutgers University Press.
Aulard, F. (1885) *L'Eloquence Parlementaire pendant la révolution*, 3 vols. Paris: Hachette.

---

[43] Outram (1989), 68–89.

Austin, M. (1931) *Experiences Facing Death*. London: Rider.

Bryant, M. (1978) Revolution and introspection: the appearance of the private diary in France. *European Studies Review*, **8**, 259–72.

Chateaubriand, F.-R. de (1958) *Mémoires d'outre-tombe* (ed. Maurice Levaillant and Georges Moulinier), 2 vols. Paris: Pléiade.

Cobb, E. (1977) *The Ecology of Imagination in Childhood*. New York: Columbia University Press.

Dehérain, H. (1905) Une autobiographie du baron Ramond, membre de l'Académie des Sciences. *Journal des Savants*, March, 121–9.

Dekker, R.M. (1969) Ego-documents in the Netherlands, 1500–1814. *Dutch Crossing: A Journal of Low Countries Studies*, **39**, 61–72.

Devic, J.-F.S. (1851) *Histoire de la vie et des travaux scientifiques et littéraires de J.D. Cassini, IV*. Clermont.

Dolomieu, D. (1793) Mémoire sur la constitution physique de l'Egypte. *Journal de Physique*, **32**, 41–2.

Dupont de Nemours, H.A. (ed.) (1906) *L'enfance et la jeunesse de Dupont de Nemours*. Paris.

Eagleton, T. (1982) *The Rape of Clarissa: Writing, Sexuality, and Class Struggle in Samuel Richardson*. Oxford: Oxford University Press.

Fisch, M.H. and Bergin, T.G. (eds.) (1944) *The Autobiography of Giambattista Vico*. Ithaca: Cornell University Press.

Fried, M. (1980) *Absorption and Theatricality: Painting and Beholder in the Age of Diderot*. Berkeley: University of California Press.

Geiger, S.N.G. (1986) Women's life-histories: method and content. *Signs*, **11**, 334–51.

Hahn, R. (1975) L'autobiographie de Lacépède retrouvée. *Dix-huitième siècle*, **7**, 49–85.

Ladner, G.B. (1967) *Homo viator*: medieval ideas on alienation and order. *Speculum*, **32**, 233–59.

Levesque, P.-C. (1807) *Notice historique sur Le Grand d'Aussy . . . lu à la séance publique du 15 messidor an X*. In Legrand d'Aussy, *Vie d'Apollonius de Tyane*. Paris: Collin.

Niggl, G. (1977) *Geschichte der deutschen Autobiographie im 18. Jahrhundert: Theoretische Grundlegung und literarische Entfaltung*. Stuttgart: Metzler.

Nussbaum, F.A. (1989) *The Autobiographical Subject: Gender and Ideology in Eighteenth Century England*. Baltimore: Johns Hopkins University Press.

Outram, D. (1978) The language of natural power: the funeral *éloges* of Georges Cuvier. *History of Science*, **16**, 153–78.

Outram, D. (1980) Politics and vocation: French science, 1793–1830. *British Journal for the History of Science*, **13**, 27–43.

Outram, D. (1983) The ordeal of vocation: The Paris Academy of Sciences and the Terror. *History of Science*, **21**, 252–73.

Outram, D. (1984) *Georges Cuvier: Science, Vocation and Authority in Post-revolutionary France*. Manchester: Manchester University Press.

Outram, D. (1987) Before objectivity: women, wives and cultural reproduction in early nineteenth-century French science', in Abir-Am and Outram (1987), pp. 17–35.

Outram, D. (1989) *The Body and the French Revolution: Sex, Class, and Political Culture*. New Haven: Yale University Press.

Perroud, C. (ed.) (1900–1901) *Correspondence de Mme Roland*. Paris: Imprimerie National.

Perroud, C. (ed.) (1905) *Mémoires de Mme Roland: Nouvelle Edition Critique*. Paris: Imprimerie National.

Roodenburg, H.W. (1985) The autobiography of Isabella de Moerloose: sex, childbearing, and popular belief in seventeenth-century holland. *Journal of Social History*, **18**, 517–40.

Rousseau, J-J. (1964) [1762] *Emile ou de l'Education* (ed. François and Pierre Richard). Paris: Garnier Frères.

Sennett, R. (1974) *The Fall of Public Man*. Cambridge: Cambridge University Press.

Silverman, S.M. (1974) Parental loss and the scientist. *Science Studies*, **18**, 259–64.

Sonntag, O. (1974) 'The motivations of the scientist: the self-image of Albrecht von Haller, *Isis*, **55**, 336–51.

Tatin, J.J. (1985) Relation de l'actualité, reflexion politique, et culte des grandes hommes dans les almanachs de 1760 à 1793. *Annales historiques de la Révolution Française*, **57**, 3–16.

Terres, J.K. (ed.) (1961) *Discovery: Great Moments in the Lives of Outstanding Naturalists*. Philadelphia: Lippincott.

Theis, L. (1976) Saints sans famille? Quelques remarques sur la famille dans le monde franc à travers les sources hagiographiques. *Revue historique*, **255**, 3–20.

Thoreau, H. (1862) Walking. *The Atlantic Monthly*, **9**, 659–67.

Trahard, P. (1936) *La sensibilité revolutionnaire*. Paris; Boivin.

Tulard, J. (1971) *Bibliographie critique des mémoires sur le consulat et l'Empire*. Geneva; Droz.

Vergnaud, M. (1945–55) Un savant pendant la révolution. *Cahiers internationaux de sociologie*, **17–18**, 123–39.

Walzer, M. (1974) *Regicide and Revolution: Speeches at the Trial of Louis XVI*. Cambridge: Cambridge University Press.

Zacher, R.A. (1976) *Pilgrimage and Curiosity*. Baltimore: Johns Hopkins University Press.

Scientific autobiographies of the French Revolution: a preliminary listing of printed and manuscript materials

Adanson, Michel, 1727–1806
MS. autobiography 17 June 1775, Library of the Institut de France Fonds Cuvier 3185/6.
Histoire Naturel du Sénégal (Paris, 1757), pp. 1–90 relate life to 1754.
Familles des plantes (1763), I, clvii–viii.

Ampère, André-Marie, 1775–1836
Mme C. Cheuvreux (ed.), *André-Marie Ampère: Journal et Correspondance 1793–1804* (Paris, 1869).
Anon., 'L'autobiographie d'Ampère'. *Bulletin de la Société des Amis de André-Marie Ampère'* (1955), **16**, 1–27.

Bonnet, Charles, 1720–1793
Raymond Savioz (ed.), *Mémoires autobiographiques de Charles Bonnet de Genève* (Paris, Vrin, 1948).

Bory de St-Vincent, Jean Baptiste, 1780–1828
Mémoires sur les cent-jours pour servir d'introduction aux souvenirs de toute ma vie (Paris, 1938).

Bosc d'Antic, Louis-Augustin Guillaume, 1759–1828
MS. autobiographie, n.d. Library of the Institut de France Fonds Cuvier 3157/1. Copy by Mme Cuvier.
A fragment of MS autobiography in Library of Muséum National d'Histoire Naturelle Paris, probably written before 1815, with a postscript after 1815.
This fragment published in Claude Perroud (ed.), *Mémoires de Mme Roland, nouvelle édition critique* (2 vols, Paris, 1905), II, 450–61.
The Fonds Cuvier version published by Claude Perroud, 'Le roman d'un Girondin'. *Revue du dix-huitième siècle*, 2 (1914), 232–57, 3 (1915–16), 348–67.

Bouvard, Alexis, 1767–1843
Charles Philippe, 'Notice sur l'astronome Bouvard'. *Revue savoisienne*, 34 (1893), 152–68, 214–25, 285–301. Fragments of autobiography.

Brisson, Mathurin-Jacques, 1723–1806
MS 'Principaux étapes de la vie de M. Brisson, écrits par lui-même'. Library of the Institut de France, Fonds Delambre, 2041/90.

Broussonet, Pierre Marie Auguste, 1761–1807.
MS. written 10 messidor au III/28 June 1795: his life from 1789: Archives Nationales, Paris, F^7 5142, fol. 45.

Chaptal, Jean Antoine, 1756–1832
Mes souvenirs sur Napoleón (Paris, 1893), 2 vols. Vol. I is Chaptal's record of his own life to 1804; vol. II deals with his recollections of Napoleon.

Cuvier, Georges, 1769–1882
MS. 2598(3) Library of the Institut de France, Fonds Flourens. 'Mémoires pour servir à celui qui fera mon éloge: écrits au crayon dans ma voiture pendant mes courses en 1822 et 1823: cependant les dates sont prises sur des pieces authentiques'. Many deletions by Mme Cuvier.
Printed in Pierre Flourens, *Recueil des éloges historiques* (Paris, 1856), 2 vols. I, 169–93.

Dupont de Nemours, Pierre Samuel, 1739–1817
L'enfance et la jeunesse de Dupont de Nemours racontés par lui-meme (Paris: Plon-Nourrit, 1906).

Fourcroy, Antoine François, 1755–1809
'Note autobiographique', personal possession of D. Duveen. Quoted in G. Kersaint, 'Antoine-François Fourcroy: 1755–1809: sa vie et son oeuvre'. *Mémoires du Muséum National d'Histoire Naturelle*, série D, 2 (1966), 1–296, passim; p. 242 discusses provenance of the MS.

Hassenfratz, Jean Henri, 1755–1827
G. Laurent, 'Une mémoire historique du chimiste Hassenfratz'. *Annales historiques de la révolution française I* (1924), 163–64. Written 1794.

Haüy, René-Just, 1743–18'
MS. 'Etat de services' (1809). Library of the University of Paris (Sorbonne) MS 1643.
Printed in A. Lacroix, 'La vie et l'oeuvre de l'Abbé René-Just Haüy. *Bulletin de la Société française de minéralogie*, 67 (1944), 15–16.

LaCépède, Bernard Germain Etienne de la Ville, Comte de, 1756–1825

MS. Fonds Goswin de Stassart, Académie royale des sciences de Belgique. Completed in 1815.
This version printed in R. Hahn, 'L'autobiographie de Lacépède retrouvée'. *Dix-huitième siècle*, 7 (1975), 49–85.
Another version in the Fonds Cuvier, 3209, Library of the Institut de France, utilised for an éloge: Georges Cuvier, *Eloges historiques* (Paris and Strasbourg, 1819–23), 3 vols. III, 285 comments on the text.

Lalande, Joseph-Jérome, 1732–1807
Extracts from MS. autobiography quoted in Comtesse de Salm, 'Eloge historique de M. de Lalande'. *Magazin Encyclopédique*, 2 (1810), 282–325.

Lavoisier, Antoine-Laurent, 1743–1794
MS. 'Notice de ce que Lavoisier, cy-devant commissaire de la Trésorerie nationale, de la ci-devant Académie des sciences, membre du bureau de Consultation des arts et métiers, cultivateur dans le district de Blois . . . a fait pour la révolution'. Printed in E. Grimaux, *Lavoisier* (Paris, 1888), 387–88.

Montucla, Jean-Etienne, 1725–99
Autobiographical letter in G. Sarton, 'Documents nouveaux concernant Lagrange'. *Revue d'histoire des sciences*, 3 (1950–1951), 129–32, dated 21 November 1794, in Archives Nationales, Paris, D. XXXVIII.IV.62.

Moreau de la Sarthe, Louis-Jacques, 1771–1826
Pierre Delaunay, 'Moreau de la Sarthe et ses souvenirs'. *Société française de l'histoire de la médecine*, 30 (1936), 353–62; 31 (1937), 13–42.

Périer, Jacques-Constantin, 1742–1818
MS. Autobiography, Bibliothèque historique de la ville de Paris, Nouvelle acquisition 147 f. 465.

Ramond, Louis François Elizabeth de Carbonnières, 1755–1827
H. Dehérain (ed.), 'Une autobiographie de Baron Ramond'. *Journal des savants*, March 1905, 121–29. Written in February 1827. The MS. is Fonds Cuvier, 3154 Library of the Institut de France.

Sage, Balthasar Georges, 1740–1824
Notice autobiographique, Paris 1818, published by the author.

St-Pierre, Jacques Henri Bernadin de, 1737–1814
Lt-Col. Largemain, 'Bernadin de St-Pierre'. *Revue de l'histoire littéraire de la France* 12 (1905), 668–92, reprints autobiography.

Tenon, Jacques-René, 1724–1816
MS. Autobiography in Fonds Cuvier 3196, Library of the Institut de France.

Valmont de Bomare, Jacques Christophe, d. 1807.
E.T. Hamy, 'Une autobiographie inédite de Valmont de Bomare'. *Bulletin du Muséum d'Histoire Naturelle* 12 (1906), 4–7.

3

From science to wisdom: Humphry Davy's life

DAVID KNIGHT

Creative science is a game for the young. Those excel in it who retain a child-like curiosity about the world down to an age when most of their contemporaries have got interested in other things like sex, power and money. While politicians, historians and playwrights (whose jobs depend upon understanding people) improve like claret with age, scientists may go off. Those engaged in scientific biography, therefore, face in particularly acute form the problem of dealing with a drama which comes to a climax early on, and then tails off. This makes for a poor read.

Davy's work on laughing gas was done when he was twenty-one; his electrochemical researches led to his discovery of potassium when he was twenty-nine; by his middle thirties he had elucidated the nature of chlorine, and invented the safety lamp for coal miners. If we concentrate upon his life in science as a matter of making discoveries which are still of importance in our own day, then his later life will have little interest for us. This is the approach in Harold Hartley's biography,[1] which is excellent when dealing with the scientific discoveries but where Davy's later years are briefly dismissed. At forty-one, in 1820, he was elected President of the Royal Society; in early retirement from 1827 he wrote dialogues about fishing and then about life in general; and he died abroad, at Geneva, in 1829 after travelling in fruitless search of health. A biography in which all this is an anticlimax is somehow defective as a work of art: we ought to impose order on life, whether it is our own or someone else's, so that it makes sense as a whole; and its interest must not therefore die before its subject does.

Davy died relatively young, at fifty; and yet in his case the problem is compounded by the change in his reputation among his contemporaries over his

---

[1] Hartley (1966), ch. 10.

last decade. Richard Holmes reflects of Coleridge[2] how differently we would have seen him had he died in 1804 when he set out hopelessly for Malta (but cheered on by a splendid valediction from Davy). Had Davy died in 1819 at forty soon after being awarded a baronetcy for his safety lamp, there would have been tremendous sadness at his early demise. He would have seemed a genius cut off at the height of his powers. He had enemies, among those who smarted at his reformation of chemical theory, and who supported the claims of George Stephenson[3] to have invented the safety lamp, seeing Davy as an arrogant metropolitan; but they were few and unimportant. His reputation as the Newton of Chemistry and the apostle of applied science stood extremely high even if (or perhaps because) he had made chemistry seem something depending on genius, superb apparatus and technique rather than accessible to all[4] – unlike engineering, in which Samuel Smiles seems to suggest that anyone could emulate his heroes through strength of character and organised common sense.

Then in 1820 Sir Joseph Banks died.[5] He had been President of the Royal Society since just before Davy was born; from being an immensely attractive young man who had sailed with Cook to Tahiti and botanised at Botany Bay, he had become an unpopular autocrat. Being President of the Royal Society gave wonderful opportunities for influencing people, but not for making friends: though 1993, being Banks' 250th anniversary, has led us to a new and clearer view of his achievements, with a number of conferences and publications. Davy speedily declared his candidature and as the inventor of the safety lamp he was unstoppable; Banks had written him a magnificent letter in 1815:

> Much as, by the more brilliant discoveries you have made, the reputation of the Royal Society has been exalted in the scientific world, I am of the opinion that the solid and effective reputation of that body will be more advanced among our cotemporaries of all ranks by your present discovery, than it has been by all the rest. To have come forward when called upon, because no one else could discover means of defending society from a tremendous scourge of humanity, and to have, by the application of enlightened philosophy, found the means of providing a certain precautionary measure effectual to guard mankind for the future against this alarming

---

[2] Holmes (1989), 362–4.
[3] Smiles (1975), ch. 6.
[4] Golinski (1992), ch. 5.
[5] Carter (1988), parts 2 and 3; Banks (1994).

and increasing evil, cannot fail to recommend the discoverer to much public gratitude, and to place the Royal Society in a more popular point of view than all the abstruse discoveries beyond the understanding of unlearned people.[6]

Great but impossible things were hoped for from Davy's Presidency, for he brought enormous scientific distinction to the post, and was not hostile as Banks had been to specialised societies: he saw them as complementary rather than as threatening to the Royal Society. But Banks was a landed gentleman and Oxford graduate; and Davy was not. His father had been a woodcarver in Penzance, and his mother had for a time kept a shop. He had been an apothecary's apprentice, and had dropped out to work with Thomas Beddoes in a medically dubious institution where gases were administered to the sick. It was possible in Regency England to rise dramatically, and social mobility is the key to Davy's life. Sir Thomas Lawrence, who painted Davy's portrait (complete with lamp) similarly rose from plebeian origins to be President of the Royal Academy; but social mobility has its price, and it was difficult for Davy to maintain his authority. It had not been easy for Banks, who had faced down a major revolt in 1783–4;[7] had Davy survived longer he might like Banks have brought the Society into line, achieved his programme of cautious reform, and manoeuvred between the hostile camps[8] which beset him: but in the autumn of 1826 his health failed. He seems to have had a stroke; and travel abroad did not work a cure. In 1827 he resigned.

To himself and to others his Presidency had been a disappointment. Although he had been involved with Stamford Raffles in setting up the London Zoo, had established good relations with specialised societies, had arranged through Robert Peel for annual Royal Medals to be awarded for distinguished science, and had begun the transformation of the Society from a club into an academy, his bad temper and what were seen as attempts to domineer were notorious. In particular, he was taken aback in his attempt to prevent corrosion of the copper bottoms of warships by attaching lumps of a more reactive metal to them. He had been one of those responsible for establishing 'applied science' rather than trial and error as the best route to technical progress. In this case, the principle of what is now called cathodic protection was sound, and in the laboratory the scheme worked very well; but in practice weeds and marine organisms adhered so strongly to the protected copper that the ships' sailing was adversely affected. He had taken

[6] Davy, J. (1858), 208.
[7] Carter (1988), ch. 9.
[8] Miller (1983), 1–48.

the whole investigation upon himself, rather than refer it to a committee (though he was assisted by Faraday,[9] who was thus introduced to electro-chemistry). To see the great Sir Humphry's theorising go adrift gave a lot of pleasure to plain men; he had expected plaudits, and he over-reacted to jokes and criticisms. His reign was neither happy nor glorious.

Historians of science now have much less trouble than they used to do in coping with those who take up institutional responsibilities though their research falls away thereby. In our day, eminent scientists may do their research largely by deputy – through research students and assistants in their laboratory – and thus maintain their profile; but this was not possible in Davy's world. He had no research school like the French were developing. Davy was keen to accept responsibility, or ambitious for power; he realised what the implications were, and about 1821 he wrote a poem about eagles teaching their young to fly up towards the Sun, the important lines of which go:

> Their memory left a type, and a desire;
> So should I wish towards the light to rise,
> Instructing younger spirits to aspire
> Where I could never reach amidst the skies,
> And joy below to see them lifted higher,
> Seeking the light of purest glory's prize.[10]

The poem, like much of Davy's heartfelt writing, is in part a lament for lost youth; and the wish expressed here was not easy for him to fulfil, as the unhappy turn of his relationship with Faraday was to show. He was happiest when he could have undisturbed at least his intermittent bursts of research activity; and conscious that in middle age his shaping spirit of imagination was no longer as potent as it had been: that there would be younger spirits going higher.

Davy was not perhaps really prepared for the problems and frustrations (as well as opportunities) involved in running science rather than doing it; and as President he was not much missed. Indeed for 1826–7 he was unable to fulfil his duties at the Royal Society, and his colleagues must have got used to his absence. It is probably significant that his Presidential Chair was filled by Davies Gilbert (formerly Giddy), a Cornish MP and Davy's first patron; a public figure perhaps but with no significant scientific research to his credit.

[9] Faraday (1991–), 330–6.
[10] Davy J. (1836), ii, 157.

The 1820s were a difficult decade for reformers in science as in politics; and indeed after Gilbert the Royal Society elected the Duke of Sussex in preference to Sir John Herschel,[11] thereby perhaps freeing him to do physics rather than administration. It is said that the best Popes have not been saints; the experiment of having a man of genius to fill Banks' place was felt to have failed, and the clock was turned back. Davy himself had hoped for Robert Peel as his successor: 'He has wealth and influence, and has no scientific glory to awaken jealousy, and may be useful by his parliamentary talents to men of science';[12] indicating that he shared the view of his colleagues about his own presidency.

Within science itself there had, however, been dramatic change with Oersted's discovery of electromagnetism; and although in 1826 Davy gave the Bakerian Lecture to the Royal Society, placing his own past and recent electrical work in context, when he died in 1829 his researches were all assimilated or superseded and he did not leave a great gap. Within chemistry, analysis seemed the most important field, rather to the disgust of Faraday and J.B. Daniell who were interested like Davy in the powers that modify matter, and in explaining chemical affinity, and not just in accurate recipes. Faraday's reputation was not yet sufficient to be comparable to Davy's, so he was in no way eclipsed by his former pupil; but he must have seemed like a survivor from an earlier epoch rather than a man dying before his time. A Romantic genius is anyway not a very useful role model. Analytical chemistry was by contrast relatively straightforward, a normal science in which a career was a reasonable aspiration for a talented person: one puzzle would lead readily on to another in a life of steady usefulness.

Davy had been one of the first men of science who could be described as a professional; earning his living by research and lecturing at the Royal Institution, and giving up the prospect of a life in medical practice. By 1829 a career in science in Britain had become much more of a possibility; particularly with the rise of formal medical education, which required lecturers who would make their reputation by research and move up a ladder of promotion; but Davy did not belong to this pattern. Indeed to him as to his generation, a career in science seemed a bit puzzling. It was not obvious that a lifetime spent adding to knowledge of Nature was well spent. That was why Davy was so delighted to have proved Bacon right, in that experiments of light done with flames did indeed lead to experiments of fruit with the safety

---

[11] Hall (1984).
[12] Davy, J. (1858), 288.

lamp: one ought not to be reclusive, but rather seek to be useful. In a sense, one should outgrow that childish curiosity that impels scientific research: and applied science made recondite theorising and experimenting respectable. But Davy also (like most of his contemporaries) saw the world in terms of natural theology; science properly understood led to wisdom and not only to knowledge. He wrote:

> Oh most magnificent and noble Nature!
> Have I not worshipped thee with such a love
> As never mortal man before displayed?
> Adored thee in thy majesty of visible creation,
> And searched into thy hidden and mysterious ways
> As Poet, as Philosopher, as Sage?

During his working life he had been a philosopher (chemical or natural), and a spare-time poet; his illness gave him time before he died to be a sage.

If the biographer has a problem with a life that falls away from a peak, his subject will have had it first and more seriously. This makes it worth our wrestling with the question of how Davy saw his life himself: and while we do not have to agree with his estimate, as far as we can reconstruct it, it is very important for our understanding. We find that whereas Davy was thrown into rage and despair by the frustrations of office and then by the illness that removed him from it, he recovered on his travels although it was clear that his sickness was mortal. He was pleased to leave his *Consolations in Travel* as a legacy to the world; and he did not in the end see his last decade as a time of failure and decline. As Georges Cuvier put it in his obituary,[13] he returned to the sweet dreams and sublime thoughts which had enchanted him in youth; and he was a dying Plato, ending an examined life in the expectation of a better and more intellectual one to come. He managed to make sense of his life, making it come full circle; and it is worth following his lead, and seeing what it meant for Davy to try to become a sage.

On 27 September 1827 he wrote gloomily in his journal:

> As I have so often alluded to the possibility of my dying suddenly, I
> think it right to mention that I am too intense a believer in the
> Supreme Intelligence, and have too strong a faith in the optimism
> of the universe, ever to accelerate my dissolution . . . I have been,
> and am taking a care of my health which I fear it is not worth; but

---

[13] Cuvier (nd), 354.

which, hoping it may please Providence to preserve me for wise purposes, I think my *duty* – G.O.O.O.[14]

The G.O.O.O. was a form of pious ejaculation often used by Davy in this
journal. Had he died at this point, his life would have been incomplete; as it
was, he had time to complete his odyssey, writing dialogues which reveal his
thoughts and feelings.

The first, *Salmonia*, was modelled on Isaak Walton's *Compleat Angler* but
dealt with the more gentlemanly sport of fly fishing, of which Davy was
extremely fond. As well as discussion of different kinds of fish, and flies with
which to tempt them on to the hook, there are topographical descriptions
giving local colour; and the dialogue form inherited from Walton allowed for
reflections on general topics and some (but not very much) variation in tone.
The book was flatteringly reviewed by Walter Scott, and Davy set about
revising it for a second edition in which he expanded the passages dealing
with life in general. In particular, he expanded a purple passage which is
clearly autobiographical:

> Ah! could I recover any thing like that freshness of mind, which I
> possessed at twenty-five, and which like the dew of the dawning
> morning, covered all objects and nourished all things that grew,
> and in which they were more beautiful even than in mid-day sun-
> shine, – what would I not give! – All that I have gained in an active
> and not unprofitable life. How well I remember that delightful
> season, when, full of power, I sought for power in others; and
> power was sympathy, and sympathy power; – when the dead and
> the unknown, the great of other ages and distant places, were
> made, by the force of the imagination, my companions and
> friends; – when every voice seemed one of praise and love; when
> every flower had the bloom and odour of the rose; and every spray
> or plant seemed either the poet's laurel, or the civic oak – which
> appeared to offer themselves as wreaths to adorn my throbbing
> brow.[15]

He also began on a further series of dialogues, beginning with a vision in the
Colosseum, which would deal with life and death, time, progress and science. These were dictated to John James Tobin, a medical student (and son

---

[14] Davy, J. (1836), ii, 281.
[15] Davy (1832), 325.

of an old friend) who was his companion on his last journey (his wife Jane was suffering ill health in London). They spent the summer months in Austria and what is now Slovenia, and the winter in Italy.

On 23 February 1829 Davy in Rome had another stroke. He dictated a letter to his brother:

> I am dying from a severe attack of palsy, which has seized the
> whole of the body, with the exception of the intellectual organ . . .
> I bless God that I have been able to finish my intellectual labours.
> I have composed six dialogues, and yesterday finished the last of
> them. There is one copy in five small volumes complete, and Mr
> Tobin is now making another copy, in case of accident to that. I
> hope you will have the goodness to see these works published.[16]

To his wife, he also sent the message that:

> I should not take so much interest in these works, did I not believe
> that they contain certain truths which cannot be recovered if they
> are lost, and which I am convinced will be extremely useful both to
> the moral and intellectual world. I may be mistaken in this point;
> yet it is the conviction of a man perfectly sane in all the intellectual
> faculties, and looking to futurity with the prophetic aspirations
> belonging to the last moments of existence.[17]

His brother and wife rushed to his bedside; he rallied, and they were bringing him home to England when he died in Geneva. The dialogues were indeed published in 1830,[18] in an elegant little volume, as *Consolations in Travel*; later editions were embellished with engravings, some by Lady Murchison, wife of the geologist and friend of Davy. The book went on selling past the middle of the nineteenth century, and was translated into Spanish, French, German and Swedish; there were also American editions. It must be called a success although there is little evidence of its being actually read very much.

The book is interesting for its references to Papina,[19] the blue-eyed and pink-cheeked daughter of an innkeeper in Ljubljana; but while perhaps his conversation with her was intellectual and refined, and she 'made some days

---

[16] Davy, J. (1836), ii, 346.
[17] Ibid., 384.
[18] Fullmer (1969), 98–9.
[19] Knight (1992a), 169–71.

of my life more agreeable than I had any right to hope' (as he told his wife),
it seems to have chiefly been scenery and philosophy rather than company
which consoled him. He enjoyed following rivers up to their sources, even
when in poor health; with fishing and shooting, at which he seems to have
been very skilled. In poetry[20] he compared human life to the course of a river.

When it reaches the sea, the water of a river is mingled with it and its
individuality is lost. Davy did not believe this of human life; despite a period
of enthusiastic materialism when with Beddoes at Clifton, he soon reverted
to a belief in the personal immortality of the better part of man. It was this
idea which in the end was to provide him with the consolations of philosophy. The central message of the first dialogue, from which the book grew,
is that when our machinery is worn out, and our work here done, we die and
migrate to another planet where we shall inhabit a more ethereal body, and
live a higher and more intellectual life. Tobin recorded that it was Davy's

> pleasure and delight during his mornings at Ischl, and when he was
> not engaged in his favourite pursuit of fishing, to work upon this
> foundation, and to build up a tale, alike redundant with highly
> beautiful imagery, fine thoughts, and philosophical ideas.[21]

On his fiftieth birthday, Davy himself wrote to his wife describing what he
was about in composing the dialogues:

> I lead the life of a solitary. I go into the Campagna to look for game,
> and work at home at my dialogues on alternate days. I hope I shall
> finish something worth publishing before the winter is over. This
> day, my birth-day, I finish my half century. Whether the work I am
> now employed on will be my last, I know not; but I am sure, in one
> respect it will be *my best*; for its object is to display and vindicate
> *the instinct or feeling of religion*. No philosopher, I am sure, *will*
> quarrel with it; and no Christian *ought* to quarrel with it.[22]

The first dialogue is not, however, very close to orthodox Christianity. One
of the lessons of the book is that death is necessary for birth or rebirth;
another that the laws of chemistry and physics do not alone govern life. The
progress described in the vision is attributed to men of genius, 'a few

[20] Ibid., 178.
[21] Tobin (1832), 120.
[22] Davy, J. (1858), 307.

superior minds'; it is not usually among the upper classes that these 'benefactors of mankind' are to be found, and they received little reward for their
activities:

> The works of the most illustrious names were little valued at the
> times when they were produced, and their authors either despised
> or neglected; and great, indeed, must have been the pure and
> abstract pleasure resulting from the exertion of intellectual superior-
> ity and the discovery of truth and the bestowing benefits and bless-
> ings upon society, which induced men to sacrifice all their common
> enjoyments and all their privileges as citizens, to these exertions.[23]

Davy thus saw himself as a martyr to science, rather than as just unpopular;
and in a great tradition of benefactors. We may find it implausible that one
so famous should consider himself to have given up all his enjoyments, but
for Davy it must have been comforting, and made sense of the unpleasant
features of his presidency.

The characters in the dialogue then met upon Vesuvius to discuss the
vision: all three, Philalethes (lover of truth) Ambrosio (immortal), a liberal
Roman Catholic, and the sceptical Onuphrio (a hermit, patron saint of a
church in Rome; whose name is a version of Humphry) must in different
degrees represent Davy himself. They discuss the relations between religious
and scientific belief, with particular reference to immortality (always a very
important matter for Davy), and to the age of the Earth. Davy's religion was
strongly personal, and he belonged to no particular church; he was surprisingly sympathetic to Roman Catholics, and rejoiced when Catholic Emanci-
pation was passed. He had little patience with William Paley's clockwork
universe, though he shared his love of fishing: he saw God as inscrutable,
and submission and trust as essential for the freshness of mind essential to
guide the wave-tossed mariner to his home. He refused to construct a system
of geology upon Genesis, but urged a progressive development of the Earth
over a long period; Charles Lyell was duly provoked to rebuke the dead Davy
for this in his *Principles of Geology*.[24]

Davy had referred to a time when 'the dead and the unknown, the great
of other ages and distant places, were made, by the force of the imagination,
my companions and friends';[25] and indeed the participants in the dialogue

---

[23] Davy, (1830), 20–1, 30, 35 (quotation), 57, 228.
[24] Lyell (1830–3), i, 144–5.
[25] Davy (1832), 325.

are joined by a mysterious figure called the Unknown. He seems to be another manifestation of Davy. Here and in a later dialogue on chemistry, he denounces materialism, and urges that chemistry is progressive and useful, and is moreover the fundamental science, wrestling with the fundamental nature of matter: it is also rather dangerous. What may have begun as a recruiting document for chemists was reworked for *Consolations* into an *apologia pro vita sua*: the chemist's life is not merely worthy, but adventurous; and scientific ambition is the highest kind, unlike that of the lawyer or the politician. The chemist both comes to understand God's world, but is also enabled to benefit his fellow men.

Davy had thus come to terms with his coming death, and with his worldly impotence. At the end, he found that his real pleasures were simple, though he loved receiving letters from the mighty, and was indignant that he had never been made a Privy Councillor as Banks was. Unlike Faraday, he had pushed social mobility to its limits; Faraday knew when to stop, remaining one of Nature's gentlemen. Faraday's religion[26] was very different from Davy's, for he was rooted in a small sect: this gave him an intense social life, but led to his eschewing worldly position and thus responsibilities. His life was almost monastic compared with Davy's, and he did not have to find out about worldly glory the hard way. Contemporaries could revere him as a kind of scientific saint, inimitable by ordinary folk; a different sort of genius from Davy's.

In the West as in China, a certain longevity is probably essential to the sage. At least he ought to survive long enough to enjoy this new found status; he may be benevolent or curmudgeonly, but should be surrounded in his declining years by disciples. Coleridge did not die in Malta, and indeed went on to outlive Davy; becoming (under the careful management of Dr Gilman) the Sage of Highgate. Thomas Carlyle similarly saw the publication of his *French Revolution* which brought him the status of a sage.[27] John Herschel returned from South Africa to assume a somewhat similar status, writing poetry and articles for the *Reviews*;[28] it was possible for a man of science to be a sage. Davy however died in the moment of victory, just as he had finished dictating *Consolations*. Had he been brought home and lived for a few more years, he might indeed have blossomed in his new role of wise man: not probably a happy one, for the man Tobin describes was grumpy.

[26] Cantor (1991), ch 4.
[27] Rosenberg (1985), part 2.
[28] King-Hele (1992), 115ff (by M.B. Hall).

Perhaps then Davy died at the right moment, though his reputation was therefore insecure: it may have been wise to die just at the point of acquiring wisdom, and avoid the risk of losing it; and perhaps inscrutable Providence was on Davy's side at the end.

Bibliography

Banks, R.E.R. *et al* (1994) *Sir Joseph Banks: a Global Perspective*. London: Royal Botanic Gardens, Kew.

Brock, W.H. (1992) *The Fontana History of Chemistry*. London: Fontana.

Cantor, G. (1991) *Michael Faraday: Sandemanian and Scientist*. London: Macmillan.

Carter, H.B. (1988) *Sir Joseph Banks, 1743–1820*. London: British Museum (Natural History).

Cuvier, G. (nd) *Eloges Historiques*. Paris: E. Durocq.

Davy, H. (1832) *Salmonia, or Days of Fly Fishing*, 3rd edition. London: John Murray.

Davy, H. (1830) *Consolations in Travel, or the Last Days of a Philosopher*. London: John Murray.

Davy, J. (1836) *Memoirs of the Life of Sir Humphry Davy*, 2 vols. London: Longman, Rees, Orme, Brown Green and Longman.

Davy, J. (1858) *Fragmentary Remains, Literary and Scientific, of Sir Humphry Davy*. London: John Churchill.

Faraday, M. (1991–) *Correspondence*. ed F.A.J.L. James. London: Institution of Electrical Engineers.

Fullmer, J.Z. (1969) *Sir Humphry Davy's Published Works*. Cambridge, MA: Harvard University Press.

Fullmer, J.Z. (1980) Humphry Davy, Reformer. In S. Forgan (ed.) *Science and the Sons of Genius: Studies on Humphry Davy*. London: Science Reviews, pp. 59–94.

Golinski, J. (1992) *Science as Public Culture: Chemistry and Enlightenment in Britain, 1760–1820*. Cambridge: Cambridge University Press.

Hall, M.B. (1984) *All Scientists Now*. Cambridge: Cambridge University Press.

Hartley, H. (1966) *Humphry Davy*. London: Nelson.

Holmes, R. (1989) *Coleridge: Early Visions*. London: Hodder and Stoughton.

King-Hele, D.G. (ed.) (1992) *John Herschel 1792–1871: a Bicentennial Commemoration*. London: Royal Society.

Knight, D. (1992a) *Humphry Davy: Science and Power*. Oxford: Blackwell.

Knight, D. (1992b) *Ideas in Chemistry: a History of the Science*. London: Athlone, and New Brunswick: Rutgers University Press.

Lyell, C. (1830–1833) *Principles of Geology*, 3 vols. London: John Murray.

Miller, D.P. (1983) Between hostile camps: Sir Humphry Davy's presidency of the Royal Society of London, 1820–1827. *British Journal for the History of Science*, 16, 1–48.

Paris, J.A. (1831) *The Life of Sir Humphry Davy*. London: Colburn and Bentley.

Rosenberg, J.D. (1985) *Carlyle and the Burden of History*. Cambridge, MA: Harvard University Press.

Smiles, S. (1975) [1874] *The Lives of George and Robert Stephenson*. Introduction by E. de Maré. London: Folio Society.

Tobin, J.J. (1832) *Journal of a Tour Made in the Years 1828–1829 through Styria, Carniola, and Italy, whilst Accompanying the Late Sir Humphry Davy*. London: W.S. Orr.

4

Robert Boyle and the dilemma of biography in the age of the Scientific Revolution

MICHAEL HUNTER

This is the story of a life of Robert Boyle that was never written, with reflections on the significance of this paradox for our understanding both of Boyle and of the nature of scientific biography in his day. In fact, of course, we have two biographical accounts of Boyle dating from the sixty years following his death, on which almost all subsequent writings on him have been based. One was the sermon preached at Boyle's funeral on 7 January 1692 by his friend, Gilbert Burnet, Bishop of Salisbury, which included a lengthy section described at the time as a 'Panegyric' of Boyle.[1] As one might expect of a funeral sermon – a genre in which Burnet specialised – this is an unashamed eulogy of Boyle, with something in common with the various verse 'elogies' of the deceased scientist that were produced in the aftermath of his death.[2] In it, Burnet drew on his intimate knowledge of Boyle (and on biographical notes that had been vouchsafed him) to paint a memorable but rather one-sided view of the great natural philosopher as a truly good man, pious, sober, modest, painstaking and intellectually innovative. His aim was to illustrate 'to how vast a Sublimity the Christian Religion can raise a mind, that does both throughly believe it, and is entirely governed by it'.[3] In many ways, this image has dominated perceptions of Boyle ever since.

In addition, we have the *Life* of Boyle by the cleric and antiquary, Thomas Birch, which was published both as the introduction to Birch's 1744 edition of *The Works of the Honourable Robert Boyle*, and as a separate book. This

[1] Burnet (1692); Hunter (1994), 90.
[2] For a list of these, see Fulton (1961), 172–4. For comparable sermons see Burnet (1682b); Burnet (1691); Burnet (1694).
[3] Burnet (1692), 21, printed in Hunter (1994), 46.

is almost as pedestrian and diffuse as Burnet's funeral sermon is brilliantly succinct, largely because of the extent to which it is made up of a series of documents concerning Boyle stitched together in chronological order. It began by publishing for the first time Boyle's autobiographical 'Account of Philaretus during his Minority' and his father, the Great Earl of Cork's 'True Remembrances', continuing with a selection of Boyle's letters and other relevant materials in an essentially annalistic vein. As such, the *Life* was a product of the antiquarian tendency in early eighteenth-century England, characterised by a proneness to pile up increasing quantities of undigested documentation with little attempt to interpret it.[4] Indeed, for his overall evaluation of Boyle – divulged in the peroration that concluded the work – Birch was predominantly indebted to Burnet's sermon, which he supplemented concerning the scientific side of Boyle's achievement with material gleaned from the introduction to the epitome of Boyle's writings issued by the scientific writer, Peter Shaw, in 1725.

Yet this left a gap. The century during which these works appeared was also the century in which modern biography was born, and the proponents of this approach were all too aware that neither a panegyric like Burnet's nor a compilation like Birch's truly represented a 'life'. As James Boswell, the author who represents the culmination of the new, frank trend in biography, wrote of his classic account of Samuel Johnson: 'he will be seen as he really was; for I profess to write, not his panegyrick, which must be all praise, but his Life; which, great and good as he was, must not be supposed to be entirely perfect'.[5] He thus unconsciously echoed the pioneer of this tradition in Boyle's own generation, John Aubrey, who considered that 'the Offices of a Panegyrist, & Historian, are much different', and whose biographical collections were consciously intended to show that 'the best of men are but men at the best'.[6] As for a compilation like Birch's, though Aubrey would have acknowledged its value in preserving materials, it was equally apparent to those who thought more than superficially about biographical writing that this, too, was not really a 'life'. The point was made most clearly by the lawyer and author, Roger North, in the only full exposition of the theory of biography to be written in England during this period, the 'General Preface' to his *Lives of the Norths* (which, unlike the biographies it prefaced, which were published in the eighteenth century, was lost from sight until 1962 and

[4] See Stauffer (1941), 248ff.; Lipking (1970), ch. 3; Korshin (1974), 515ff.
[5] Boswell (1934), i. 30. Cf. ibid., iii, 155.
[6] Hunter (1975), 79; Aubrey (1898), i, 11–12.

printed in full only in 1984).[7] In North's view, useful though it might be to 'make gatherings and excerpts out of letters, books, or reports' concerning a man, 'those are memorials, or rather bundles of uncemented materials, but not the life'.[8]

Writing the biography of a man like Boyle presented a particular challenge in that (though Burnet's sermon had celebrated him almost entirely as a great Christian, referring only briefly to his intellectual activity), what was required was the life of a thinker. An attempt needed to be made to understand a man's mind and to chart the growth of his ideas, in other words, to write an intellectual biography, a genre as yet in its infancy.[9] Aubrey was at least aware of the desirability of doing justice to such considerations, as shown particularly in his lengthy life of Thomas Hobbes, while a pioneering example had been set by Pierre Gassendi in his *Life of Peiresc*, translated into English in 1657 and often cited at this time as a model biography.[10] Gassendi specifically justified writing the life of an intellectual as against a man of action – the staple of most biographical writing up to that point – and, particularly in his last book, he attempted to give a view of Peiresc's intellectual characteristics.[11] Though some attempts were made along these lines during the period under consideration, however – the earliest of them an abortive life of Boyle which will be dealt with here – it was only in the late eighteenth century, with works like Johnson's *Lives of the Poets* and Robert Anderson's *Life of Johnson*, that this genre began to come into its own.[12]

At least to some extent, Birch was aware of the shortcomings of his attempt to provide a satisfactory biographical treatment of Boyle. 'The only qualities I can engage for are industry and fidelity', he wrote in his preface, apologising that he had written the book 'from the best materials that could be procured at this distance of time, and without most of those advantages which Dr *Burnet* bishop of *Salisbury*, and Dr *William Wotton*, who had the same design near fifty years ago, might have obtained for the execution of it'.[13] It is, indeed, true that Burnet and Wotton had intended to write a

---

[7] Clifford (1962), 27–37; North (1984), 49–86.

[8] North (1984), 77.

[9] Cf. Korshin (1974). For an example of the life of an intellectual which makes almost no attempt to deal with his ideas, see Isaak Walton, 'The Life of Mr Richard Hooker' in Walton (1927), 153–249.

[10] Aubrey (1898), i. 321–403; Gassendi (1657). This is cited, for instance, in Burnet (1682a), sig. A5. See also below p. 118.

[11] See Gassendi (1657), sig. a2v and passim; Joy (1987), esp. 53–4.

[12] Lipking (1970), ch. 13; Korshin (1972–3).

[13] Birch (1772), i, iii.

proper life of Boyle to supplement Burnet's sketch in his funeral sermon. Let us, therefore, look at the abortive projects to which Birch there refers, to see what was intended, and what might have been achieved had circumstances permitted.

Gilbert Burnet and formal biography

In his sermon, Burnet had explicitly stated that he was 'reserving to more leisure and better opportunities, a farther and fuller account' of Boyle.[14] Moreover, the auguries for this seemed good. Burnet was one of the leading exponents of biographical writing in late Stuart England, author of accomplished lives of such figures as the Earl of Rochester, Sir Matthew Hale, William Bedell and the Dukes of Hamilton and Castleherald; there was also a strong biographical element in the posthumously published *History of My Own Time* on which he was already at work at this time. It is not surprising that Boyle's friend, Sir Peter Pett, was confident in his expectation that a life of Boyle by Burnet would materialise, and he explicitly compared Burnet's task with Gassendi's in his *Life of Peiresc*. To assist Burnet in his task, Pett prepared extensive notes intended to supplement the generalised eulogy of the funeral sermon with greater detail. These show that he took it for granted that the putative life should not eschew 'negative' aspects of Boyle's career – for instance, books criticising him – though he expected a degree of discretion on Burnet's part as to exactly what should and should not be included. He also presumed that Burnet would consult and perhaps publish documentation concerning Boyle, as he had in certain of the lives he had already published. Indeed, in one of these, he had specifically apologised for inserting 'most of the Papers at their full length', on the grounds that thus greater verisimilitude might be achieved and bias avoided.[15]

In the event, Burnet never completed his intended life. Perhaps he was overwhelmed by the sheer amount of manuscript material available, which Pett assured him 'would aske some moneths time to peruse', a problem intensified by the sheer pressure of business on one of the leading ecclesiastical politicians of his day: it is worth noting that all of Burnet's published lives date from before his elevation to the episcopate. In addition, Burnet may

---

[14] Burnet (1692), 22, printed in Hunter (1994), 47.
[15] Burnet (1677), sigs. a1v, a5; see also Burnet (1685), which advertises this component through the inclusion of a list of letters (sig. b4). For Pett's notes see Hunter (1994), 58–83, where they are published for the first time; see also ibid., xxxii–v.

have come to feel that he had little to add to the evaluation of Boyle that he had already given, and in his *History of My Own Time* he noted of his funeral sermon that 'I gave his character so truly that I do not think it necessary now to enlarge more upon it'.[16] If he *had* completed the life that he had promised, however, what would it have been like?

Burnet's extant lives represent the quintessence of what has been described as 'the theory of formal biography' which flourished in late seventeenth-century England.[17] He had a clear view of the value of biographical writing, expressed in the prefaces to his books. Convinced that 'No part of History is more instructive and delighting, than the Lives of great and worthy Men', he thought it important to give an account which was at once truthful and inspirational.[18] His prescription included an element of detail: thus his life of the lawyer, Sir Matthew Hale, told of Hale's solicitude for old horses and dogs, and it gave instances to illustrate Hale's virtues, for example his fairness to Quakers; Burnet also quoted notes by Hale whose roughness showed 'they were only intended for his Privacies'.[19] But there was a limit to the amount of detail that he saw as appropriate. In the preface to *The Life and Death of Sir Matthew Hale*, Burnet criticised other biographers for 'writing Lives too jejunely, swelling them up with trifling accounts of the Childhood and Education, and the domestick or private affairs of those persons of whom they Write, in which the World is little concerned'.[20]

In many ways, a life like that of Hale may be seen as a long version of the kind of evaluation that Burnet gave in his funeral sermon for Boyle — essentially one-sided, an 'unqualified eulogy', in the words of one commentator on the account of Hale.[21] Burnet's 'Design in Writing' was 'to propose a Pattern of Heroick Virtue to the World', avoiding 'saying any thing of him, but what may afford the Reader some profitable Instruction'.[22] The result is an almost monotonous hymn of praise. Wherever Burnet sensed a difficulty, he elided it. Thus, of Hale's writings on natural philosophy, in which he took issue with the views of Boyle and Henry More, Burnet tactfully noted how he displayed 'as much Subtilty in the Reasoning he builds on them, as these Principles to which he adhered could bear', finding something to praise in

---

[16] Burnet (1897–1900), i, 344.
[17] Stauffer (1930), 253.
[18] Burnet (1682a), sig. A3.
[19] Ibid., 19, 138ff., 164–5.
[20] Ibid., sig. A7.
[21] Calamy (1829), i, 327n.
[22] Burnet (1682a), sig. b1v, 11. Cf. sig. A7v, 118.

the virtuosity that allowed Hale to apply himself to such matters at all.[23] He also omitted all reference to Hale's controversial role in the witch trial over which he presided at Bury St Edmunds in 1664, and said little about his somewhat iconoclastic views on the need for law reform, which might have revealed him as less at one with the establishment than the book otherwise suggested.[24]

The limitations of *The Life and Death of Sir Matthew Hale* were most cogently expressed by Roger North. North had no fondness for Burnet, whose Whig politics were fundamentally opposed to his own (his animus may have been further fuelled by Burnet's opposition to the North family's ally, the Duke of Lauderdale).[25] But, though this may account for the vigour of North's attack on Burnet, and for some of his more politically motivated accusations against Hale himself, it does not devalue his viewpoint. This is epitomised by North's opinion of Burnet's lives both of Lord Rochester and of Hale: 'the persons of whom none ever knew, but must also know that those written lives of them are mere froth, whipped up to serve a turn'.[26]

North was especially incensed by the life of Hale, who, as a leading judge, was one of the principal colleagues of North's brother, Lord Keeper Guildford, and who therefore plays a significant subordinate role in North's own biographical writings. In North's view, for all Hale's positive virtues, these were balanced by weaknesses that Burnet simply ignored. North acknowledged of Hale that 'in the main he was a most excellent person, and in the way of English justice an incomparable magistrate', but he was at pains to illustrate how 'he had his frailties, defects, prejudices, and vanities, as well as excellencies'.[27] In particular, North considered that 'he was the most flatterable creature that ever was known', and 'it is most certain his vanity was excessive; which grew out of a self-conversation and being little abroad'. North illustrated these failings at length, dwelling on the partiality in Hale's judicial rulings which in his view resulted from them, and claiming that they meant that 'he would be also a profound philosopher, naturalist, poet, and divine, and measured his abilities in all these by the scale of his learning in the law which he knew how to value'.[28]

---

[23] Ibid., 25–6. For a brief account of Hale's writings, see Shapin and Schaffer (1985), 222–4.

[24] See Heward (1972), esp. chs. 7, 15. For hints of Hale's reformist view of the law, see Burnet (1682a), 121, 176.

[25] See North (1984), 167. Cf. ibid., 26–7; Burnet (1897–1900), i. 184–5.

[26] Ibid., 77. Cf. North (1890), iii, 102.

[27] North (1890), iii, 101, i, 91.

[28] Ibid., i, 81, 83. See also ibid., i, 79ff., iii, 93ff., passim.

North's view of Hale encapsulates his view of biographical writing as a whole, which was no less didactic than Burnet's, but which took for granted that such ends were best served by verisimilitude: 'it is the office of a just writer of the characters of men to give every one his due, and no more'.[29] He thus echoed the views of Aubrey and of Gassendi in his *Life of Peiresc* – who argued the value of divulging even 'such as they themselves would not willingly have the World acquainted with', on the grounds that this was likeliest to 'discover a man, and shew his inside' – in contrast to the attitude of Burnet and similar biographers, who saw such verisimilitude as somehow demeaning.[30] Moreover, North exemplified this in his own lives – perhaps particularly that of his brother, John – in which the failings of his subjects are sympathetically treated, and in which telling anecdotes are used to enhance the overall picture in a manner worthy of Aubrey.[31]

What is particularly interesting in the case of Hale is that North considered that Burnet was giving voice to an accepted view of the judge as an icon of his age, which he therefore had a duty to set right. 'I have understood it absolutely necessary for me', he wrote, 'to show Hales in a truer light than when the age did not allow such freedom, but accounted it a delirium or malignancy at least, not to idolize him: and thereby to manifest . . . that he was not a very touchstone of law, probity, justice, and public spirit, as in his own time he was accounted'.[32] Moreover, one can extrapolate from this to a broader point that North made in his 'General Preface' concerning one of the drawbacks of such biographical writing as had been carried out prior to his time, that it was often calculated to serve party interest: 'Some have wrote lives purely for favour to certain theses, opinions, or sects; and then all is in an hurry to come at them'.[33]

This may seem a lengthy digression from the subject of Boyle. However, it is possible to extrapolate directly from the state of affairs concerning Hale to that concerning Boyle, both from the point of view of the degree of verisimilitude that seemed appropriate in writing about him, and of the underlying pressure towards a kind of idolatry. For, though North unfortunately devoted only a few lines to Boyle, he said just enough to indicate his belief that a similar state of affairs existed in relation to Boyle as to Hale. Moreover, Hale and Boyle had a great deal in common, despite the fact that Hale

[29] Ibid., i, 91. Cf. North (1984), 60ff.
[30] Gassendi (1657), sig. a2. Compare Stauffer (1930), 252–3. For Aubrey, see above, p. 116.
[31] North (1984), 25ff. and 87–162.
[32] North (1890), i, 90–1.
[33] North (1984), 63.

attacked Boyle's pneumatics in his treatises on natural philosophy which have already been referred to. Both were widely respected and pious laymen, and it is not surprising that, noting Boyle's death in his memoirs, the nonconformist Edmund Calamy explicitly linked him and Hale together as 'the two great ornaments of King Charles [II]'s Reign'.[34]

North is, in fact, one of the few contemporary commentators on Boyle who gives any indication that he considered the great man to be anything less than perfect. In his autobiography, àpropos his regimen for his own health, North recorded an occasion when, in the company of the former royal physician, William Denton, he 'was fleering at the infinite scrupulosity Mr Boyle used about preserving his health'. In this connection, he instanced Boyle's extreme fear of coming into contact with anyone who had had contact with victims of smallpox – an anxiety on Boyle's part that is also documented in other sources[35] – while he sarcastically noted how Boyle 'had chemical cordials calculated to the nature of all vapours that the several winds bring', of which he partook according to the direction of the wind on any particular day. Evidently he felt that Boyle exemplified a valetudinarianism similar to that to which he gave critical if sympathetic coverage in his life of his brother John. What is most interesting, however, is that he added a crucial aside which shows that he saw Boyle as the beneficiary of the same uniformly sympathetic press as was the case with Hale. To quote North's exact words: 'But Mr Boyle had such a party, that all he did was wise and ingenious, and the doctor took his part and defended his regimen'.[36]

Such evidence as survives suggests that this was all too true, and that Boyle represented an exemplar for a certain 'sect' (as North would have put it), in this case the emergent new science. By the late seventeenth century, the Scientific Revolution badly needed a pantheon, and Boyle was elevated to this even during his lifetime. Thus his merits and achievements were consciously championed by Henry Oldenburg, both in his *Philosophical Transactions* – in which Boyle was regularly promoted and his books prominently and enthusiastically reviewed – and in Joseph Glanvill's *Plus Ultra* (1668). The latter work was encouraged by Oldenburg to compensate for what he saw as the deficiencies of Thomas Sprat's *History of the Royal Society* from the point of view of giving a detailed account of what the Royal Society and its leading members had actually achieved. In this case, Oldenburg's sedulity

---

[34] Calamy (1829), i, 327. Note also that both appear in Anon (1792), 15–32, 45–70.
[35] See the note by Caspar Lindenberg printed in Maddison (1969), 221–2.
[36] North (1890), iii, 146. Cf. North (1984), 139–42.

was such that he even solicited information from Boyle about his unpublished writings – including the part of his *Usefulness of Experimental Natural Philosophy* which at that point remained unfinished – so that it could be incorporated verbatim into a highly adulatory account of Boyle as a man 'who alone hath done enough to oblige all Mankind, and to erect an *eternal Monument* to his *Memory*. So that had this *great Person* lived in those days, when men *Godded* their *Benefactors*, he could not have miss'd one of the first places among their *deified Mortals*'.[37] Dedications of books to Boyle carry a similar message, of a steady stream of adulation for him as 'The Worthy Patron and Example Of all Vertue'.[38]

Hence, like Hale, Boyle was a kind of icon for significant groups in late Stuart England; even if people felt reservations like North's they were unlikely to divulge them. It is thus revealing that there is evidence that, privately, Bishop Burnet himself was not entirely uncritical of Boyle. In his autobiography, unpublished in full until this century, Burnet's account of Boyle is mainly as adulatory as his funeral sermon; significantly, however, a reservation creeps in, where he notes that Boyle 'was perhaps too eager in the pursute of knowledge'. In the reworked version of this that was published as part of his *History of My Own Time*, this was omitted, as with other character sketches which Burnet revised 'due to the desire of deprecating censure, whether deserved or undeserved', part of a general pursuit of what was politically prudent for the public version of his book.[39] As far as Boyle was concerned, only fashionable wits like Samuel Butler and disenchanted Tories like Jonathan Swift felt able to give voice to a critical view of the great man.[40]

William Wotton and the challenge of contextualism

This makes it all the more appropriate that the person who inherited Burnet's mantle was William Wotton, a protegé of Burnet's who had won notoriety by taking sides in the 'Battle of the Books' in his *Reflections upon Ancient and Modern Learning* (1694), a defence of the moderns against the advocacy of the ancients by Swift's mentor, Sir William Temple.[41] This had included

---

[37] Glanvill (1668), 92ff., esp. 92–3, 104–6, the latter taken verbatim from Royal Society Boyle Papers 8, fol. 1, which is endorsed 'Receaved from Mr Boyle April 25th 1666. by me, Henry Oldenburg': the significance of this document will be fully assessed in the forthcoming 'Pickering Masters' *Works of Robert Boyle*. See also Shapin (1991), 298–9.

[38] Fulton (1961), 160. See also ibid., 155ff., passim.

[39] Foxcroft (1902), xviii, 464–5; cf. Burnet (1899–1900), i, 343–4. See also Firth (1938), 189ff.

[40] Swift (1939), 239–40; Butler (1759), i, 404–10. See also Shapin (1991), 310.

[41] See Levine (1991), ch. 1; Hall (1948).

an adulatory account of Boyle as one of the leading vindicators of the new philosophy; indeed, Boyle was one of the moderns who received most references in Wotton's book as a whole.[42] Wotton was a child prodigy who had obtained a Cambridge B.A. at the age of twelve and a half and become F.R.S. at twenty-one, and John Evelyn, who was associated with Wotton's project from an early stage, was ecstatic that this 'universaly learned, and indeede extraordinary person' had taken on the project of writing the life of 'our Hero, to whom there are many Trophes due'. 'Mee thinks', he wrote, 'I already see my noble Friend Mr *Boyle*, Rising againe, and made Immortal by Mr *Wotton*, & Mr *Wotton* by Mr *Boyle*'.[43]

Compared with Burnet, Wotton was at an immediate disadvantage in trying to write a biography of the kind that Roger North considered appropriate. North agreed with many others, including Samuel Johnson, that 'They only who live with a man can write his life with any genuine exactness and discrimination', and that anyone lacking such personal knowledge – as was the case with Wotton and Boyle – was severely handicapped.[44] Wotton did what he could to offset this. In addition to inheriting the biographical materials with which Boyle had provided Burnet, together with the notes by Pett which have already been referred to, he also sought personal reminiscences from men who had known Boyle: as a result, he received detailed letters from Evelyn, Archdeacon Thomas Dent and the Revd James Kirkwood which gave significant information of which Birch was later to make selective use.[45]

By way of compensation for his lack of personal acquaintance with his subject, however, Wotton obtained access to Boyle's papers, and systematically worked through these.[46] As was made clear by the rather premature advertisement for his book that appeared in the *London Gazette* in 1699, what Wotton evidently planned was a fully documented life complemented by a selection 'from his Manuscripts and Papers of Experiments never Published, and also from a vast number of Letters to and from the most Learned Men in Europe, his Correspondents', the latter evidently intended to enhance Boyle's reputation by illustrating the acclaim in which he had been held by the international scholarly community.[47] At least in part, it is clear

---

[42] Wotton (1697), 262; Hall (1948), esp. 1055, 1061. See also Wotton (1697), 194–6, 205–6, 260, 371.

[43] Pepys (1932), 241; Hunter (1994), 86; British Library Evelyn Papers MS 39, no. 792.

[44] Boswell (1934), ii, 446. Cf. North (1984), 77, 80; Evelyn (1850–2), iii, 359–60.

[45] See Hunter (1994) for a full account of this material.

[46] Hunter (1992), xii–xiii.

[47] Quoted in Maddison (1957), 93. For background, see Goldgar (1995), ch. 3.

that Wotton was aspiring to something new: his aim was to contextualise and evaluate Boyle's findings, and his book, or at least the surviving fragment of it, represents what is evidently the first attempt at an intellectual biography in a modern sense in England. As he explained to Sir Hans Sloane in 1709, 'I do not barely design to give an Account of what Mr Boyle had done already, (for that is what every body already know's,) but to compare his Inventions & Discovery's with the Discovery's & Inventions that have bin made since'.[48]

Unfortunately, Wotton's project also failed to materialise, despite the fact that he applied himself to it at intervals over a period of some years.[49] He undertook the task in 1696 and was certainly at work on it later in that decade. Despite the fact that 'The discouragements I met with since I undertook it were so many, that I have often wished that I had let it alone, or never thought of it', a further campaign of work occurred in 1702–3, as is shown by his correspondence with Evelyn at this time.[50] Lastly, we know that he was studying Boyle's pneumatical experiments in 1709, when he wrote to Sir Hans Sloane requesting the loan of books by such predecessors of Boyle's as Galileo and Otto von Guericke. He may have been stimulated to work on this theme at this point by the publication in that year of Francis Hauksbee the elder's *Physico-Mechanical Experiments on Various Subjects*. Though much of a chapter on Boyle's pneumatics which was probably written in that year survives, however, the book was never published and the rest of the manuscript has disappeared.

To some extent, the reasons for Wotton's failure to complete the work are obscure. A biography of Wotton by his son-in-law, William Clarke, states that the papers which it comprised 'were unhappily either lost or destroyd, & he was so much affected by this Misfortune, to have spent so much Time to no Purpose, that he had not Resolution enough to think of turning all the same Books & papers over a second Time, & beginning again': it appears that the loss of the papers was linked to Wotton's hasty absconding from his Buckinghamshire parish in 1714 due to a scandal.[51] But there were also difficulties about the task that Wotton had set himself, illustrating that, for all his significance as a pioneer in intellectual biography, this was a genre which as yet needed further refinement. 'This is a wide field, &

[48] British Library MS Sloane 4041, fol. 317, quoted in Hunter (1994), li.
[49] Hunter (1994), xxxvi–xlvi.
[50] Evelyn (1850–2), iii, 388–9; Hunter (1994), xli–iv.
[51] Quoted in Hunter (1994), xlv–vi. See also Bayle (1734–41), x, 206; Levine (1991), 404.

I am extreamly fearfull that I shall miss my way in it', he told Sloane in the letter already cited, and some of his dilemmas are apparent from the extant chapter.[52] This opens with quite a perceptive account of the context of Boyle's work on pneumatics, including his aims and precursors, together with a description of the air-pump itself. Thereafter, however, the text settles down to a detailed account of the content of Boyle's *New Experiments Physico-Mechanical, Touching the Spring of the Air and its Effects* (1660), section by section, interspersed by notes on ancillary works by Boyle and by information about research on related problems done since, particularly by the elder Hauksbee, whose book on related topics had, as we have seen, appeared in 1709.

Here, two interrelated problems about Wotton's task become apparent. For one thing, in his concern with the growth of knowledge, Wotton tended to minimise the attention that he gave to controversies which he regarded as closed by the work of Boyle and his successors: thus the controversy with Hobbes and Linus was here ignored, despite the fact that Wotton had specifically alluded to it in his earlier *Reflections*, while Boyle's reflections on the significance of his findings were severely curtailed.[53] Equally telling was the attention that Wotton gave to Hauksbee, a living and active scientist, the preface to whose book boasted of his '*Great* and *Further Improvements*' to the air-pump that Boyle had pioneered. Initially, in implicit criticism of such claims, Wotton made a point of emphasising that, since Boyle was the originator of the pump, he deserved some of the credit for later achievements. Thereafter, however, Wotton made repeated flattering remarks about Hauksbee, 'whose Discoverys of this sort are the latest we have', usually along such lines as: 'This Experiment . . . has bin much farther carried by Mr Hauksbee, in whose Hands every Thing of this Nature receives Improvement'.[54] Inevitably, this had the effect of 'writing down' Boyle.

Indeed, Wotton's chapter represents an early instance of a more general problem about Boyle's scientific work in the eighteenth century, in that its importance was acknowledged, but it tended to be seen as superseded by more up-to-date research. The matter was put most candidly by a commentator of 1808:

> I have not the slightest intention to undervalue the *acknowledged* importance of Mr Boyle's physical discoveries; considering the state

---

[52] Printed from the MS in British Library Add. MS 4229 in Hunter (1994), 111–48.
[53] Wotton (1697), 196; Hunter (1994), lii–iv. See also Shapin and Schaffer (1985).
[54] Hauksbee (1709), sig. alv; Hunter (1994), 120–1, 126. Cf. ibid., 134–5, 143–4, 146–8.

of science at the time he lived, they are unquestionably highly credit-
able to his industry, and ingenuity, *and are the steps by which the
subsequent philosophers have ascended.* But it is perfectly well
known, that the modern discoveries have extremely reduced the
value of all ancient speculations in natural philosophy, have indeed
placed them very much upon the footing of exploded errors.[55]

In Boyle's case, matters were made worse by the extent to which he was
overshadowed by a figure who had built on his findings and gone far towards
transcending them, namely Isaac Newton, and particularly his *Opticks*, published in 1704. This is seen even in Wotton's chapter, since, in relation to
Experiment 31 of the *New Experiments*, he cited both Hauksbee and John
Keill in invoking attraction as a partial explanation of the adhesion of marble
plates which Boyle had attributed solely to the power of the vacuum, adding:
'But the Discovery of that general Law of Matter was not then known, that
was reserved to the incomparable Sir Isaac Newton to find out'. More generally, the way in which the publication of the *Opticks* had the effect of down-
grading Boyle is apparent in the contrast between the two volumes of John
Harris' scientific encyclopaedia, *Lexicon Technicum* for Boyle figures much
more prominently in the first, published in 1704, than he does in the second,
which came out in 1710.[56]

This tendency is equally apparent in Peter Shaw's 1725 epitome of
Boyle's writings, despite the fact that the very publication of that book bore
witness to an interest in Boyle's scientific work, while Shaw's introductory
notes were full of praise for him. For Shaw's footnotes repeatedly refer to
the way in which Boyle's discoveries had been furthered and enhanced by
Newton's, 'whose words we can never use too much', as where he writes:
'This doctrine cannot be better illustrated, confirm'd, and improv'd, than by
the words of that incomparable philosopher Sir *Isaac Newton*'.[57] Up to a
point, there might have been an element of 'political correctness' in this
during Newton's lifetime, since undue praise for Boyle might have been seen
as an implicit criticism of his eminent successor.[58] But the trend continued

---

[55] Weyland (1808), viin.

[56] Harris (1704, 1710), e.g. s.v. Air, Corpuscles, Particles, Pores. For the Wotton quotation, see
Hunter (1994), 145, where Wotton's sources are identified.

[57] Shaw (1725), i, 223, 386, and passim. See also Golinski (1983), though in this instance Shaw's
notes have more of a Newtonian flavour than he implies (p. 25); other more recent writings are
also cited, but Newton is predominant, as Shaw in effect acknowledges: Shaw (1725), i, iv.

[58] All the more so in view of the use of Boyle by Leibniz and other continental natural philosophers
as a weapon against Newton: see Clericuzio (1990), 561–2.

after 1727, as can be seen in later scientific encyclopaedias such as that of Ephraim Chambers or the *Encyclopaedia Britannica*, in which Boyle's discoveries are acknowledged, but much more space is devoted to the findings and prescriptions of his younger contemporary.[59]

The result was that people tended to lose interest in the detailed content of Boyle's work, as against valuing a generalised memory of him as a great and good man. In the words of William Clarke, the cleric and antiquary who inherited Wotton's papers after his 'Life' was abandoned: 'For my part, I have no ambition to become an author, and especially upon these subjects: the examining *cosmical qualities*, and weighing of *igneous corpuscles*, are things I have no great taste for; though I have a great value for Mr Boyle's memory, as a true Philosopher and Christian'.[60] In particular, stress was laid on Boyle's deep religiosity: this was all the more crucial because of the long-standing suspicion that science encouraged irreligious tendencies in contemporary society, which Boyle's example could be conclusively used to refute.[61] Indeed, it is even possible that this became a more necessary function for Boyle to play as Newton's science took centre stage, due to the extent to which rumours about Newton's questionable orthodoxy circulated.[62] It may thus be significant that in a tract entitled *Remarks on the Religious Sentiments of Learned and Eminent Laymen*, published in 1792, eighteen pages were devoted to Boyle but only five to Newton, one of which comprised a transcript of the inscription on his tomb.[63]

Birch, Miles and the burden of adulation

In these circumstances, the chances of an honest and balanced biographical account of Boyle being written dwindled, and the problems are well illustrated by what we know about the making of Birch's *Life*. For, despite all the profuse documentation that this incorporated, it was highly selective of biographical material which was available to Birch in manuscript – including quite revealing documents, some of them bequeathed by Boyle himself – which might have helped to give the rounded view of the man to which a biographer like Roger North would have aspired.

---

[59] See Chambers (1741), passim, e.g. s.v. Air, Corpuscular Philosophy, Vacuum; *Encyclopaedia Britannica* (1797–1802), passim, e.g. iv, 377.
[60] Nichols (1812–15), iv, 454.
[61] See Hunter (1990a).
[62] Westfall (1980), 829.
[63] Anon. (1792), 10–32.

Thus we know that at the time when Birch was writing his book a good deal of material survived his relating to among Boyle's papers interest in casuistry, in other words, the salving of his conscience on difficult moral issues on which he sought the advice of leading churchmen; this was a characteristic facet of the heightened religiosity of the sixteenth and seventeenth centuries which was not to the taste of the more settled Anglicanism of the eighteenth century. Sir Peter Pett in his biographical notes specifically stated that he understood that 'many' of Boyle's papers comprised 'Bishop [Thomas] Barlows & Mr [Richard] Baxters resolutions of cases of conscience put to them by Mr Boyle', while Thomas Dent in his letter to Wotton assured him that such documents would provide 'very proper materialls for your historicall account'.[64] Certainly the one such item that *does* survive among Boyle's papers – a report on his confessional interviews with Burnet and with Bishop Edward Stillingfleet in the last year of his life – gives an extraordinary insight into Boyle's preoccupations and priorities, as I have made clear in my commentary on the published text of it.[65] Otherwise, however, the documentation of Boyle's casuistical concerns that the archive once contained has disappeared. Moreover, this was evidently due to conscious choice on the part either of Birch, or of his collaborator, the dissenting minister Henry Miles, who provided much of the material for Birch's *Life*. This is made clear by a letter from Miles to Birch in which he explained:

> NB I have some cases of consc[ience] proposed to Dr Barlow with the Drs answers in MSS. which are too long to be inserted in the life, and the resolutions being in the manner of the Schoolmen, it was thought by a very judicious friend to whom I shewd the MS better to omit em as not suited to the genius of the present age &c.[66]

Hence material which would have illustrated the authentic details of Boyle's spiritual life was passed over, and instead Birch's *Life* retailed bland commonplaces about the great man's religious outlook.

Positive censorship appears to be in evidence concerning material illustrating Boyle's attitude to magic and his interest in alchemy. One problem for those solicitous about Boyle's reputation in the early eighteenth century was that the view was abroad that he was credulous, and his protagonists felt it

---

[64] Hunter (1994), 77, 106.
[65] Hunter (1993).
[66] Quoted in Hunter (1993), 86.

important to head off such criticism. Thus Peter Shaw in his 1725 edition of Boyle refers to 'the general opinion which has crept abroad of Mr *Boyle*', that he 'easily credited uncertain accounts of things uncommon and extraordinary', citing such examples as his belief in the weapon salve and other surprising cures, or in the alchemical transmutation of metals.[67] Shaw, of course, responded that Boyle *was* careful in his enquiries, but yet *did* vindicate many 'things that should sound odd, strange, and shocking to vulgar ears'. 'This, surely, is not unphilosophical or irrational', he wrote, arguing: 'On the contrary, do not those who, without evidence, and merely from common report, believe Mr *Boyle* to have been credulous, give us a remarkable instance of their own credulity?'[68] In addition, in connection with alchemy and chemistry, Shaw specifically replied to 'some learned men' who censured Boyle 'for cultivating an art which they apprehend to be unworthy of him'. Not only, he responded, was chemistry integral to Boyle's scientific work; in addition, 'chymistry is another thing than what it formerly was; it has now made ample amends for all the darkness and obscurity wherein it formerly involved the world'.[69] Clearly, however, the suspicion was there and needed to be answered.

Birch quoted Shaw's defence of Boyle at length in his first published account of Boyle, which preceded his biography by nine years: this was one of the entries that he wrote for the extended English edition of Pierre Bayle's *General Dictionary, Historical and Critical* which he and others produced between 1734 and 1741.[70] On the other hand, presumably because he was personally intrigued by such things, Birch rather tactlessly juxtaposed this with a quite disproportionate emphasis on Boyle's alchemical interests in the article's profuse footnotes. In one he included a lengthy summary of Boyle's 1678 tract on the degradation of gold by an anti-elixir (which Shaw had dealt with by unobtrusively splicing a summary of it into his epitome of Boyle's *Usefulness of Natural Philosophy* at an appropriate point); in another he quoted extracts from *Certain Physiological Essays* to illustrate Boyle's conviction of the possibility of transmutation, while he also published for the first time the famous 1676 letter from Newton to Oldenburg commenting on Boyle's article in *Philosophical Transactions* on the incalescence of mercury.[71] In addition, in a further footnote in which he made a somewhat

[67] Shaw (1725), i, ix. See also Maddison (1969), 189, 194–5.
[68] Ibid., i, ix, xii.
[69] Ibid., iii, ccliii, cclvi, 259.
[70] Bayle (1734–41), iii, 541–60, esp. 555–6. See also Osborn (1938).
[71] Ibid., iii, 557–9; Shaw (1725), i, 78.

clumsy attempt at a survey of Boyle's natural philosophy as a whole, he gave a rather surprising prominence to Boyle's views about 'cosmical qualities'.[72]

By the time of the *Life*, however, this emphasis had been corrected. Though a handful of brief references to Boyle's alchemical concerns remained, they are buried so deep in Birch's chronological narrative as to render them almost invisible.[73] Moreover, though evidence in Boyle's writings for his alchemical interests could hardly be suppressed from the edition of his works that Birch produced, Boyle's correspondence was censored from this point of view. Though, happily, many letters to Boyle still survive, many do not, and to a disproportionate extent, letters to Boyle from alchemists were omitted from the edition, or, worse, were actually destroyed. This is revealed by surviving inventories made by Miles, which include various letters from shadowy figures on alchemical topics that are no longer extant: a case in point is the series of letters that Boyle received from John Matson of Dover, all of which have disappeared except one – perhaps retained by oversight – in which its author asks Boyle for advice on quite arcane alchemical matters.[74] Moreover, Miles' comments in the surviving inventories of these letters suggest that this destruction was due to conscious policy on his part, since a number of such items are described as 'N[o] W[orth]'.[75] The result was to reinforce a view of Boyle as a strict mechanist, wholly opposed to non-mechanical forces, which had already begun to be promoted in the early eighteenth century. Moreover the historiographical legacy of this has been an inappropriately strong contrast between Boyle's position and that of contemporary Helmontians and others, in contrast to the intermediate position which he actually adopted and which is only now being brought to light.[76]

Other letters which no longer survive are described in Miles' inventories as 'Enthusiastic', 'unintelligible', 'immaterial or useless'. There are also letters linking Boyle with Quakers, most notably the emigré, Benjamin Furly, in Amsterdam, Boyle's relations with whom would otherwise be wholly undocumented; the extant notes on these lost items reveal a shared interest

---

[72] Ibid., iii, 548–54, esp. 549–50. Cf. Henry (1994). It is interesting that Boyle's *Cosmical Qualities* and his alchemy are both given comparable prominence in the account of Boyle by John Campbell in *Biographia Britannica* (1747–66), ii, 923–7; however, on Campbell's known hermetic interests, see DNB.

[73] Birch (1772), i, civ–vi, cxi, cxxxi–ii.

[74] Royal Society Boyle Letters, iv, 42.

[75] Royal Society Boyle Papers 36, fols. 144–5, 161–2.

[76] Clericuzio (1990).

in the *Cabala Denudata*, Christian Knorr von Rosenroth's monumental cabalistic collection, published in 1677, of which Furly sent Boyle a copy, and various related matters. Again, this tidying up of the archive by its eighteenth-century custodians has had an unfortunate historiographical legacy, providing the basis for a misleading view of Boyle as unequivocally opposed to 'enthusiasts'.[77] In addition, certain letters (for instance, concerning Boyle's estates in Ireland and Dorset) were evidently destroyed because they seemed trivial, and this, too, has had a distorting effect on our image of the great man, making him seem more Olympian and detached from his context than would otherwise be the case. Though it should be noted that it is not absolutely clear when these losses occurred, it was certainly after Miles and Birch had access to the material — since it is from their notes on the lost letters that these conclusions have been derived — while it was almost certainly before Miles' widow gave the material to the Royal Society in 1769.[78]

The deliberateness of such censorship is equally clear with the biographical material that Birch had access to, including a highly revealing memorandum comprising notes dictated by Boyle to Burnet towards the end of his life. Burnet, with characteristic circumspection, made use of a handful of points from this in his funeral sermon, and this example was followed by Birch.[79] But what Birch entirely ignored was the lengthy section of the document — nearly half its overall length — which, as I have illustrated elsewhere, provides an extraordinary insight into Boyle's ambivalent attitude towards magic, even recording his reaction when he was offered the opportunity to consult spirits through a magical glass.[80] What is more, an earlier passage in the same document which revealed Boyle's belief in day fatality was the subject of a revealing comment by Miles in a letter to Birch of 21 October 1742. Evidently Wotton had mentioned this in a section of his 'Life' which Miles saw but which is now lost; Wotton seems to have believed that Boyle was anxious that the matters dealt with in this document should be included in his life, even if they reflected badly on him, though Wotton accompanied this by a note which 'dissaproves of the notion & hints of the evils which may attend such a perswasion'. Miles's view was: 'Now I woud not wish to

---

[77] Jacob (1977).

[78] It should be noted that there seems to have been a partly random element in the survival of the letters: but there can be no doubt of the more or less conscious censorship involved. I intend to give a full account of this material in a future study.

[79] See Hunter (1994), xxx–xxxi, xcv (n. 270), 47, 49, 50; Burnet (1692), 23–4, 28–9.

[80] Hunter (1990*b*), 388–91.

see this in his Life, it may have bad Consequences to some who have a Veneration for Mr B[oyle], & the omitting it, will do no injury to his Memory — I submit this to your Consideration & better Judgement'.[81]

As Miles' remark suggests, Boyle's reputation was too precious to be compromised by the biographical truthfulness to which a man like Roger North (and, in this case, evidently Wotton) aspired. Boyle remained — as he had been in his lifetime — an icon of the new science, and for this what was best was to continue retailing the view of him so brilliantly set out in Burnet's funeral sermon. Moreover, this is precisely what happened. 'We are at a loss which to admire most, his extensive knowledge, or his exalted piety. These excellencies kept pace with each other', wrote James Granger in his *Biographical History of England*, while the retrospective account of Boyle which appeared among the *éloges* of Condorcet saw his general role in encouraging scientific endeavour and his exemplary life as more important than any specific discoveries on his part.[82]

It is appropriate to end with the *éloges*, because the case of Boyle illustrates with particular clarity a more general problem of scientific biography in the eighteenth century, seen perhaps most acutely in the *éloges* of the Paris Académie des Sciences, though similar tendencies are in evidence in the Republic of Letters during this period as a whole.[83] For the polemical use of the biographical form to present an elevated view of intellectuals, magnifying their virtues and minimising their faults, had an inhibiting effect on the growth of serious biographical writing. A parallel to the state of affairs concerning Boyle is provided by the material furnished by Conduitt to Fontenelle for his *éloge* of Newton, which goes out of its way to present the most favourable possible view of the great man and to smooth off his rough edges; thus Conduitt presented Newton's chronological and chemical studies as diversions and even managed to reduce the sharp comments about Newton's religious heterodoxy with which he was furnished by John Craig to commonplace platitudes.[84] It was in the nature of *éloges* to be self-fulfilling statements of the harmony of virtue and scientific productivity.

It is small wonder that in 1800 — the age of Boswell — the *Monthly Review* should attack the Royal Society of Edinburgh for the generalised *éloges* which

[81] British Library Add. MS 4314, fol. 70.
[82] Granger (1775), iv, 84; Condorcet (1847–9), ii, 104–6. Cf. the very similar remarks by W.T. Brande in the 1824 supplement to the *Encyclopaedia Britannica*, quoted in Maddison (1969), 195.
[83] Paul (1980); Hahn (1971), ch. 2; Outram (1978); Goldgar (1995), ch. 3.
[84] Turnor (1806), 158f., esp. 163, 165; Westfall (1980), 829.

it produced on the model of the French academy, instead advocating more specific information:

> We would willingly exchange the general terms and long phrases, which seem to involve virtues and talents that were splendid and various, for distinct instances of goodness, and evident specimens of mental ability; we would resign abstract designations, in order to gain sensible images.[85]

Yet, as Richard Yeo has pointed out, when attempts were made to present a more honest, less stylised view of Newton than that of the *éloges* in the early nineteenth century, this was initially seen as denigration.[86] Only slowly was the stranglehold of adulation broken, and the history of attempts to write Boyle's life helps to illustrate the reasons why. Moreover, in Boyle's case, it is only in the late twentieth century that a sustained attempt is being made to revise the image of the scientist which the eighteenth-century tradition so successfully created.[87]

Acknowledgements

I am grateful to the editors of this volume and to Mordechai Feingold for helpful comments on a draft of this essay. I also benefitted from the discussion which followed an oral presentation of it at the University of Alabama at Birmingham in March 1993.

Bibliography

Anon. (1792) *Remarks on the Religious Sentiments of Learned and Eminent Laymen*. London.

Aubrey, J. (1898) *Brief Lives, chiefly of Contemporaries*. ed. Andrew Clark. 2 vols. Oxford: Clarendon Press.

Bayle, P. (1734–41) *A General Dictionary, Historical and Critical*. (ed. J.P. Bernard, Thomas Birch, John Lockman, etc.), 10 vols. London.

Biographia Britannica (1747–66) *Biographia Britannica: or, the Lives of the Most eminent Persons Who have flourished in Great Britain and Ireland, From the earliest Ages, down to the present Times*, ed. William Oldys and Joseph Towers. 6 vols. London.

Birch, T. (ed.) (1772) *The Works of the Honourable Robert Boyle . . . To which is prefixed The Life of the Author*, 2nd edition, 6 vols. London.

[85] Stauffer (1941), 536–7.

[86] Yeo (1988), 271. Of course, Yeo goes on to point out how nineteenth-century tastes could accommodate a more tempestuous figure. For an analogy on the case of Boyle see Wilson (1862), 233ff., which presents a view of him as a melancholy man of genius, though this seems to have had little influence.

[87] See esp. Hunter (1994), lxiii–lxxix.

Boswell, J. (1934) *Life of Johnson*. (ed. George Birkbeck Hill, rev. L.F. Powell), 6 vols. Oxford: Clarendon Press

Burnet, G. (1677) *The Memoires of the Lives and Actions of James and William Dukes of Hamilton and Castleherald &c.* London.

Burnet, G. (1680) *Some Passages of the Life and Death Of the Right Honourable John Earl of Rochester*, London.

Burnet, G. (1682a) *The Life and Death of Sir Matthew Hale, Kt.* London.

Burnet, G. (1682b) *A Sermon Preached at the Funeral of Mr James Houblon.* London.

Burnet, G. (1685) *The Life of William Bedell.* London.

Burnet, G. (1691) *A Sermon Preached at the Funeral of the Right Honourable Anne, Lady-Dowager Brook.* London.

Burnet, G. (1692) *A Sermon Preached at the Funeral of the Honourable Robert Boyle; at St Martins in the Fields, January 7, 1691/2.* London.

Burnet, G. (1694) *A Sermon Preached at the Funeral of the Most Reverend Father in God John By the Divine Providence Lord Archbishop of Canterbury.* London.

Burnet, G. (1897–1900) *History of My Own Time*. (ed. Osmund Airy), 2 vols. Oxford.

Butler, S. (1759) *The Genuine Remains in Prose and Verse*, 2 vols. London.

Calamy, E. (1829) *An Historical Account of My Own Life*. (ed. J.T. Rutt), 2 vols. London.

Chambers, E. (1741) *Cyclopaedia: or, an Universal Dictionary of Arts and Sciences*, 4th edition, 2 vols. London.

Clericuzio, A. (1990) A redefinition of Boyle's chemistry and corpuscular philosophy, *Annals of Science*. 47, 561–89.

Clifford, J. (1962) *Biography as an Art: Selected Criticism 1560–1960*. London: Oxford University Press.

Condorcet, M.J.A.N. Caritat, Marquis de (1847–9) *Oeuvres*. (ed. A. Condorcet O'Connor and M.F. Arago), 12 vols. Paris.

Encyclopaedia Britannica (1797–1802) *Encyclopaedia Britannica*, 3rd edition, 18 vols. and 2 vols. supplement. London.

Evelyn, J. (1850–2) *Diary and Correspondence of John Evelyn, F.R.S.* (ed. William Bray [and John Foster]), 4 vols. London.

Firth, Sir C. (1938) Burnet as an historian. In *Essays Historical & Literary*. Oxford: Clarendon Press, pp. 174–209

Foxcroft, H.C. (1902) *A Supplement to Burnet's History of My Own Time*. Oxford: Clarendon Press.

Fulton, J.F. (1961) *A Bibliography of the Honourable Robert Boyle*, 2nd edition. Oxford: Clarendon Press.

Gassendi, P. (1657) *The Mirrour of True Nobility & Gentility. Being the Life of the Renowned Nicolaus Claudius Fabricius, Lord of Peiresk* (Eng. trans. by William Rand). London.

Glanvill, J. (1668) *Plus Ultra: or, the Progress and Advancement of Knowledge since the Days of Aristotle.* London.

Goldgar, A. (1995) *Impolite Learning: Conduct and Community in the Republic of Letters, 1680–1750.* New Haven and London: Yale University Press.

Golinski, J.V. (1983) Peter Shaw: chemistry and communication in Augustan England. *Ambix*, 30, 19–29

Granger, J. (1775) *A Biographical History of England from Egbert the Great to the Revolution*, 2nd edition, 4 vols. London.

Hahn, R. (1971) *The Anatomy of a Scientific Institution: The Paris Academy of Sciences, 1666–1803*. Berkeley: University of California Press.

Hall, A.R. (1948) William Wotton and the history of science. *Archives Internationales d'Histoire des Sciences*, 5, 1047–62.

Harris, J. (1704, 1710) *Lexicon Technicum*, 2 vols. London.

Hauksbee, F. (1709) *Physico-Mechanical Experiments on Various Subjects*. London.

Henry, J. (1994) Boyle and cosmical qualities. In *Robert Boyle Reconsidered* (ed. Michael Hunter). Cambridge: Cambridge University Press, pp. 119–38.

Heward, E. (1972) *Matthew Hale*. London: Robert Hale.

Hunter, M. (1975) *John Aubrey and the Realm of Learning*. London: Duckworth.

Hunter, M. (1990a) Science and heterodoxy: an early modern problem reconsidered. In *Reappraisals of the Scientific Revolution* (eds. D.C. Lindberg and R.S.Westman). Cambridge: Cambridge University Press, pp. 437–60.

Hunter, M. (1990b) Alchemy, magic and moralism in the thought of Robert Boyle. *British Journal for the History of Science*, 23, 387–410.

Hunter, M. (1992) *Letters and Papers of Robert Boyle: A Guide to the Manuscripts and Microfilm*. Bethesda, MD: University Publications of America.

Hunter, M. (1993) Casuistry in action: Robert Boyle's confessional interviews with Gilbert Burnet and Edward Stillingfleet, 1691. *Journal of Ecclesiastical History*, 144, 80–98.

Hunter, M. (1994) *Robert Boyle by Himself and his Friends, with a Fragment of William Wotton's Lost 'Life of Boyle'*. London: William Pickering.

Jacob, J.R. (1977) *Robert Boyle and the English Revolution*. New York: Burt Franklin.

Joy, L.S. (1987) *Gassendi the Atomist: Advocate of History in an Age of Science*. Cambridge: Cambridge University Press.

Korshin, P. (1972–3) Robert Anderson's *Life of Johnson* and early interpretative biography. *Huntington Library Quarterly*, 36, 239–53.

Korshin, P. (1974) The development of intellectual biography in the eighteenth century. *Journal of English and Germanic Philology*, 73, 513–23.

Levine, J.M. (1991) *The Battle of the Books: History and Literature in the Augustan Age*. Ithaca: Cornell University Press.

Lipking, L. (1970) *The Ordering of the Arts in Eighteenth-century England*. Princeton: Princeton University Press.

Maddison, R.E.W. (1957) A summary of former accounts of the life and work of Robert Boyle. *Annals of Science*, 13, 90–108.

Maddison, R.E.W. (1969) *The Life of the Honourable Robert Boyle, F.R.S.* London: Taylor & Francis.

Nichols, J. (1812–5) *Literary Anecdotes of the Eighteenth Century*, 9 vols. London.

North, R. (1890) *The Lives of the Norths* (ed. Augustus Jessopp), 3 vols. London.

North, R. (1984) *General Preface and Life of Dr John North* (ed. Peter Millard). Toronto: University of Toronto Press.

Osborn, J.M. (1938), Thomas Birch and the *General Dictionary* (1734–41). *Modern Philology*, 36, 25–46.

Outram, D. (1978) The language of natural power: the 'éloges' of Georges Cuvier and the public language of nineteenth-century science. *History of Science*, 16, 153–78.

Paul, C.B. (1980) *Science and Immortality: the Eloges of the Paris Academy of Sciences (1699–1791)*. Berkeley: University of California Press.

Pepys, S. (1932) *Letters and the Second Diary* (ed. R.G. Howarth). London: J.M. Dent & Sons.

Shapin, S. (1991) 'A scholar and a gentleman': the problematic identity of the scientific practitioner in early modern England. *History of Science*, 29, 279–327.

Shapin, S. and Schaffer, S. (1985) *Leviathan and the Air-Pump: Hobbes, Boyle and the Experimental Life*. Princeton: Princeton University Press.

Shaw, P. (1725) *The Philosophical Works of the Honourable Robert Boyle, Esq.*, 3 vols. London.

Stauffer, D.A. (1930) *English Biography before 1700*. Cambridge, MA: Harvard University Press.

Stauffer, D.A. (1941) *The Art of Biography in Eighteenth-century England*. Princeton: Princeton University Press.

Swift, J. (1939) *A Tale of a Tub and other Early Works* (ed. Herbert Davis). Oxford: Basil Blackwell.

Turnor, E. (1806) *Collections for the History of the Town and Soke of Grantham. Containing Authentic Memoirs of Sir Isaac Newton*. London.

Walton, Isaak (1927) *The Lives of John Donne, Sir Henry Wotton, Richard Hooker, George Herbert & Robert Sanderson* (ed. George Saintsbury). London: Oxford University Press.

Westfall, R.S. (1980) *Never at Rest: a Biography of Isaac Newton*. Cambridge: Cambridge University Press.

Weyland, J. (1808) *The Hon. Robert Boyle's 'Occasionall Reflections'. With a Preface, &c.* London.

Wilson, G. (1862) 'Robert Boyle', in *Religio Chemici. Essays*. London, pp. 165–252.

Wotton, W. (1697) *Reflections upon Ancient and Modern Learning*, 2nd edition. London.

Yeo, R. (1988) Genius, method, and morality: images of Newton in Britain, 1760–1860. *Science in Context*, 2, 257–84.

5

Alphabetical lives: scientific biography in historical dictionaries and encyclopaedias

RICHARD YEO

Theophilus Hopkins was a moderately famous man. You can look him up
in the 1860 *Britannica*. There are three full columns about his corals and
his corallines, his anemones and starfish. It does not have anything very
useful to say about the man. It does not tell you what he was like. You
can read it three times over and never guess that he had any particular
attitude to Christmas pudding.

Peter Carey, *Oscar and Lucinda*, 1988, p.7

Today we conventionally consult encyclopaedias for biographical information. Probably at least as many readers seek this in the *Encyclopaedia Britannica* as in the *Dictionary of National Biography*, or *Who's Who*, or other
specialist biographical dictionaries. But in the famous encyclopaedias of the
Enlightenment such biographical information was absent. Neither Ephraim
Chambers's *Cyclopaedia* of 1728 nor the more renowned *Encyclopédie*
(1751–80) edited by Denis Diderot and Jean D'Alembert, contained any biographical entries: when personal names appear these usually represent some
larger intellectual doctrine, such as Baconianism, Cartesianism, or Newton-
ianism.[1] From the early nineteenth century encyclopaedias began to include
the biographical entries we now expect. However in its 1985 revision, the
Britannica decided to relegate all biographical articles from its Macropaedia
to the Micropaedia, except for those on '100 people who profoundly affected

---

[1] See for example 'Baconisme', 'Cartesianisme' and 'Newtonianisme' in Diderot and D'Alembert
(1751–80), vol. 1, 8–10, vol, 2, 716–26 and vol. 11, 122–5.

world history'. This was a departure from its previous twentieth-century editions, but still a long way from the rigorous exclusion practised by its eighteenth-century predecessors.[2]

The absence of biography in the eighteenth-century encyclopaedias was a conscious choice, in response partly to the separate genre – the 'historical' dictionary – which included biography and geography. Influential works of this kind by Louis Moreri and Pierre Bayle in France, and their English successors such as the *Biographia Britannica*, edited by William Oldys (and later by Andrew Kippis), carried entries on the lives of aristocracy, political, religious and military leaders, and also on philosophers and men of science, although usually not in any comprehensive manner.[3] We often tend to lump these encyclopaedias and dictionaries together indiscriminately as reference works, thus missing their distinctive rationales. In this essay I discuss some of the assumptions governing the initial exclusion of biography from encyclopaedias and, secondly, the implications of its eventual inclusion by the end of the eighteenth century. This requires some attention to the historical dictionaries deriving from the late seventeenth-century French publications just mentioned. I refer to important English translations (and extensions) of these but I do not attempt a systematic study of the place of scientific biography in these or in the encyclopaedias that followed them. Such an investigation is beyond the scope of this discussion. However, I believe that the point raised here about the conventions regarding biography in these works needs to be considered in any larger treatment of the representation of scientific lives in the major reference works of the period.

Where is the biography? Historical dictionaries vs encyclopaedias

The omission of biographical entries from encyclopaedias was more than a convenient division of labour. Encyclopaedias did not include biography because they were seeking to record knowledge, not lives. By the eighteenth century most of these publications were alphabetical in arrangement, but their contents continued to be influenced by some of the principles that guided the older, systematic encyclopaedic compendia, such as Johann Heinrich Alsted's *Encyclopaedia* of 1630 which recorded the major branches of

---

[2] Whiteley (1992), 85.
[3] Lipking (1970), 67 says Oldy's *Biographia Britannica* (1747–66) 'is an initial great landmark in that trail of systematic lives which would lead eventually to the *Dictionary of National Biography*'. See Retat (1991), 505 on the importance of historical dictionaries in the republic of letters in the eighteenth century.

scientia: namely, systematic knowledge embodied in the seven liberal arts of the medieval university curriculum.[4] From the 1690s a new kind of scientific and technical dictionary appeared: for example, Antoine Furetiere's *Dictionnaire Universel* (2 vols, 1690), John Harris's *Lexicon Technicum* (2 vols, 1702; 2nd edition 1710), and Chambers' *Cyclopaedia* (2 vols, 1728). These were subtitled 'dictionaries of arts and sciences', and they sought to record the major terms and concepts of these categories or fields of knowledge. However, by this time there had been shifts in the definition of *scientia*: the new experimental sciences, although not strictly fitting the axiomatic character expected in the traditional definition, were now included, together with the mechanical arts.[5] These works went through several editions and were aimed both at scholars and educated readers. For example, Chambers' *Cyclopaedia* was published by subscription in 1728 at a cost of four guineas for the two folio volumes. The first volume carried the names of 375 initial subscribers and its immediate success encouraged Chambers and his printers to begin a second edition; this was issued in 1738 and again in 1741. The range of subjects covered was broader than in Harris' *Lexicon*, but both works disseminated scientific terminology, including that of Newtonian natural philosophy, beyond scientific circles. Editions of these two English works usually ran to about 2000 copies; both were still being published and supplemented in the 1750s and hence prepared an audience and a market for the major encyclopaedias, such as the *Encyclopédie* and the *Encyclopaedia Britannica*. But the readers of these scientific dictionaries could not learn anything about individual natural philosophers. There was still no rationale for a coverage of lives, even the lives of the persons closely involved in the various arts and sciences.

The first edition of the *Encyclopaedia Britannica*, advertised to appear in 100 weekly parts from January 1768, was published in three volumes in 1771. Although differing markedly from Chambers' *Cyclopaedia* in the way it organised scientific knowledge – in large systems, rather than small entries on terms[6] – it continued the exclusion of biography. It did carry a short entry on 'Biography' which praised the moral, didactic value of the genre and remarked that there were not enough suitable lives being written for the edification of the young. But the editors felt no compulsion to remedy this

---

[4] See Alsted (1630) for the full title.
[5] Chambers (1728), vol. 1, ii for his chart of knowledge.
[6] See *Proposals for printing by subscription a work, intitled Encyclopaedia Britannica* (Edinburgh, 1768). On its plan and organisation, see Yeo (1991). For a study of the first edition, see Kafker (1994).

in their own publication. When it defined the two leading systems of natural philosophy – 'Cartesianism' and 'Newtonian Philosophy' – the *Britannica*, like Chambers', allowed no comment on the lives or character of the two individuals who founded these doctrines, although the 'Cartesians' were said to embrace a 'romantic system'. The entry on 'Newtonian philosophy' refers the reader to the one on 'Optics', where Newton is mentioned only in a description of the reflecting telescope.[7] Francis Bacon's method is praised in 'Aether', where Newton's speculation about elastic fluids is censured; Boyle is mentioned in 'Chemistry'. Even major religious leaders such as Luther, Calvin and Knox do not receive special treatment, although the first two are mentioned in entries on their doctrines. Thus eighteenth-century readers seeking a biographical account of men of science needed to consult one of the 'historical' or biographical dictionaries. How did these dictionaries treat the lives and discoveries of men of science?

Louis Moreri's *Grand Dictionnaire Historique, ou mélange curieux de l'histoire sacrée et profane*, first published in Lyon in 1674 and then issued in an expanded second edition, is usually regarded as the first reference work (other than bibliographies) to summarise a range of subjects in strictly alphabetical order.[8] Although the word 'biography' was not mentioned in the title it was the first 'modern' work to include biographical entries. It set a pattern for the 'historical dictionaries' of the eighteenth century, professing to include history, geography and genealogy as well as the lives of famous people. This biographical content was intelligible given the contemporary definition of biography as the history of a life. Although history and biography were classified as distinct genres by Francis Bacon and John Dryden, they were rarely distinguished in practice until the mid eighteenth century.[9]

---

[7] [Smellie] (1768–71), vol. 1, 555 on 'Biography'; vol. 2, 39–40; on Cartesians'; vol 3, 399 on 'Newtonian Philosophy'; vol. 3, 424 on the reflecting telescope. The absence of biography is also apparent in Scott (1765). Here there is an entry for 'Copernicus' but this describes the 'astronomical instrument invented by Mr Whiston'. I have not done a careful check of the German encyclopaedia edited by Johann Heinrich Zedler, but Collison 1964, 105 says it was unusual in having biographical entries of living persons. In the volume published in 1740 there is a one column entry on Newton which describes him as the greatest philosopher and mathematician of his time, and gives some account of his early studies. Newton's epitaph is printed under his name and the reader is referred to the larger entry on '*Newtonische Philosophie*'. See Zedler (1731–50), vol. 24, 411–12 and 413–16.

[8] The second edition appeared in two volumes in 1681. I use the English translation, from the 1692 Utrecht edition, published in two volumes in 1694. The French version reached a 24th edition in 1759. For further information, see Miller (1981); Retat (1987), 505.

[9] Bacon (1974), 72–5; Stauffer (1941), 517. In 1728 Chambers's *Cyclopaedia*, which did not include biographies, nevertheless gave the accepted definition of the art: a 'Biographer' was 'an author who writes the history, or the life of one or more persons'. Chambers (1741), vol. 1. Also Hall 1797, vol. 1: 'Biography: a species of history which records the lives and characters of remarkable persons.'

Thus in describing the 'historical' contents of his work Moreri promised that:

> besides the Lives of the Ancient and Modern Emperors, Kings and
> Princes, which are to be found in other Authors, here are the Histor-
> ies of all who have ever been Famous for Arts, Arms, or any Thing
> else; which are either not to be met with, or but very slightly
> touch'd upon in the Chronicles of Nations, etc. Here are also Lists
> of Learned Men in all Faculties, with Catalogues of their most
> Remarkable Writings.[10]

More than half of Moreri's content was biographical but many of his entries
gave the names of aristocratic figures with very little account of their lives.[11]
It is significant that as a compendium that broke with the systematic
(non-alphabetical) organisation of encyclopaedic works, Moreri's dictionary
also introduced biography — a subject dealing with contingent matters that
clashed with the axiomatic and deductive character of *scientia*. Biography,
together with geography and the mechanical arts, was a subject outside
formal university studies; to the best of my knowledge, Moreri's was the first
alphabetical reference work to include it.[12]

An English translation was published in 1694 as *The Great Historical, Geographical and Poetical Dictionary*. The editors said that Moreri 'thought himself oblig'd to gratify the Nobles and Gentry of that Kingdom [France], with
an Account of their Illustrious Families, and the Famous Exploits of their
Ancestors'. They followed his charter by 'paying the same regard to the
Nobles and Gentry of England, Scotland and Ireland', thus expanding the
content of the work. The lives of 'Philosophers' were also included but these
entries were not extensive: Aristotle got two columns, Roger Bacon ten lines,
Copernicus half a column, Kepler eleven lines, Francis Bacon half a column.
Newton was not included because he was still alive, but Robert Boyle's name

Tytler (1778–84), vol. 2, 1156: 'a species of history which records the lives and characters of
remarkable person'. On the clash between the historical and biographical material of Moreri's
dictionary and the ideal of a didactic and encyclopaedic order pursued by Diderot and D'Alembert,
see Retat (1991), 508. For a deliberate distinction between 'biographical and historical matter',
see Aiken and Enfield (1799–1815), vol. 1, 3 and the 'Discours Préliminaire' to the *Biographie
Universelle*, Paris: M. Freres, 1811, vol. 1, vii–xviii.

[10] Moreri (1694), preface, no pagination. The copy I used was bound in one volume.

[11] R. Christie (1884), 194. When starting the *Dictionary of National Biography* (1895), Leslie Stephen
still felt it necessary to say: 'I exclude names which are only names, because otherwise I should
have to publish (amongst other things) all the parish registers. A biographical dictionary should
surely consist of biographies, however brief '. R. Christie (1884), '6.

[12] Moreri's, however, was not the first compendium to include subjects outside the scholastic curriculum. Sebastian Munster's *Cosmographie* (1544) contained an even wider range of miscellaneous
information, arranged under various 'heads'. See Strauss (1966).

is mentioned under 'Boyle, Richard of Burlington', from where the reader is sent to the letter 'R' to discover one of the longest entries (two and a half columns) under 'Robert Boyle'. This entry contained hardly any description of Boyle's scientific work – apart from a list of his books – but it did present him as a pious man whose natural philosophy was compatible with 'the truth of the Christian Religion in General'. He was civil to strangers, offering charity to 'those in want', and unlike some men of learning, he was 'decently Cheerfull'. This was an early statement of the ideal Christian Philosopher or Virtuoso, following Gilbert Burnet's sermon at Boyle's funeral in 1692. But unlike later biographical accounts of natural philosophers, none of these entries in Moreri provided much discussion of the controversies in which the individuals had been involved. Brief portraits of character took precedence over intellectual details, and a list of works was appended at the end of each entry, curiously detached from the history of the life, which was largely presented in terms of personal appearance, behaviour and moral character.[13]

Despite the great success of Moreri's work, which reached a twenty-fourth edition in 1759, its reputation today derives partly from its influence on Pierre Bayle, whose famous two-volume Dictionnaire Historique et Critique of 1697 began as an attempt to remedy the errors of his compatriot. However, when Bayle wrote his Dictionnaire (an enlarged second edition appeared in 1702) he concerned himself only with the biographical, not the geographical or historical, articles.[14] This focus made it a stimulus and model for the more strictly biographical dictionaries published during the eighteenth century. But this genealogy should not be oversimplified, because while Bayle's Dictionnaire was explicitly biographical in content – that is, it listed names of people in alphabetical order – there was no clear emphasis on the 'life history' of the person; this was lost in the mass of intervening footnotes (the hallmark of Bayle's 'critical' style) which digressed into philosophical commentary. This method displayed his belief that intellectual claims needed constant reassessment. Bayle distanced himself from contemporary publications – particularly Moreri's – underlining the difference

---

[13] Moreri (1694), see under 'Copernicus', 'Bacon', 'Kepler', 'DesCartes'. Another English work based on Moreri was produced by Jeremy Collier in 1721 and extended in 1727. See Collier (1721). On Boyle's self-image and the assessments of his contemporaries, see Hunter (1990) and his chapter in this volume.

[14] The first edition of Bayle was published in Rotterdam to avoid French censorship; it reached a ninth edition in 1741. Bayle omitted names adequately covered in Moreri. For details, see Burrell (1981). For selections in English from the French edition, see Bayle (1965).

between 'other Historical Dictionaries and mine. I am not contented, according to the custom of those dictionaries, to give a general account of a man's life'.[15] Rather, he explained, his work offered comment and criticism on certain aspects of the lives and teachings of the persons covered in the *Dictionnaire*. These lives were selected to allow scope for the discussion of the topics nearest to Bayle's heart: Protestantism, religious doctrine and other philosophical issues concerning evidence and authority. There were no articles on Copernicus, Galileo, Boyle, Harvey, Leibniz or Newton. Thus although there was an entry on Kepler, and support of the heliocentric theory, the new science did not feature strongly in Bayle's work.[16]

Bayle's *Dictionnaire* appeared in English in two translations of 1709 and 1710 and these were followed by two further competing editions of 1734–38 and 1734–41. The most significant of these was a new and greatly enlarged one undertaken by Thomas Birch and a number of collaborators. This translation, *A General Dictionary, Historical and Critical*, published in ten volumes between 1734 and 1741, was in some ways a distinct work that added many new lives, especially English ones.[17] The editors explained that Bayle had chosen 'such Articles only as best suited his views, or for which he had materials already prepared, [and] he omitted a great many persons illustrious for their rank and dignity, as Emperors, Kings, Princes, etc or conspicuous for their knowledge in the Arts, the Sciences and Polite Literature.' The editors aimed to add entries on 'the most famous personages throughout the Dictionary of Mr BAYLE, and have enlarged and compleated his Articles, wherever we apprehended them to be defective'. Recognising the danger of being 'voluminous' they excluded 'whatever relates to Geography, as being foreign to a Work of this kind'.[18] This restraint did not stop the edition running to ten volumes, a very large number for that time.

The result was a more strictly biographical dictionary than Moreri's and a less critical one than Bayle's. While offering only lives, and not history or geography, it reinstated the illustrious names that Bayle had omitted when he decided on a more extensive discussion of significant figures. This also

---

[15] Bayle (1826), vol. 4, 164–5.

[16] Burrell (1981), 93–4. D'Alembert indicated that Bayle's work was not a '*dictionnaire historique*' but a philosophic and critical *dictionnaire*, where the text is only the pretext of the notes. Retat (1987), 510.

[17] See Bayle (1710); for the ten volume work edited by Birch, see, Bayle (1734–41). The rival five volume translation was Bayle (1734–38).

[18] Bayle (1734–41), vol. 1, preface. For Birch's major role in this, see Osborn (1938).

meant the addition of British names, because when Bayle had heard that the translation of Moreri was being greatly expanded he decided 'not to treat of the illustrious men of Great Britain.'[19] The English editors agreed that Bayle 'draws the characters of such persons, relates the particulars of their lives, discovers the several springs of their actions, and examines the judgement that has been, or may be formed of them'. But they suggested that the text seemed to be written merely for the sake of the notes. Nevertheless this English edition did not entirely escape the influence of Bayle's format – footnotes dominate many pages – yet in expanding the original two volumes into ten, it greatly extended the coverage of names.[20]

It is interesting that this coverage included a significant proportion of men of science from the seventeenth century. Thomas Birch contributed at least twenty new entries on both major and minor figures. Indeed, under his direction, the first volume was dedicated to Sir Hans Sloane, President of the Royal Society, and addressed its Fellows in these terms:

> Gentlemen, although you are so assiduously engaged in the most
> rational and most sublime Pursuits, those of the Mathematicks and
> of Nature, we yet presume to interrupt them a few moments . . . as
> the subject of it is the Lives of eminent Men, many of whom bear so
> near a resemblance to Your Selves.[21]

The editors also professed that 'the bare Names of NEWTON and of BOYLE raise the most exalted Ideas, and image to us something more than human'. But while the lives of natural philosophers were regarded as worthy of record, very few details of their work were given. The entry on Newton – partly derived from Fontenelle's *eloge* – is an exception in that his theory of gravity is summarised, with some of the mathematical reasoning given in the footnotes; but his work in optics is only noted in passing. Hooke's skill in experimentation is praised, but there is no description of this work or its significance.[22] Compared with Moreri's, there was slightly more integration of scientific material into the narrative of the life: publications were cited in footnotes rather than merely listed at the end of the article. The editors also published, for the first time,

---

[19] This remark occurs in the reprint of Bayle's preface to the original French edition. See Bayle (1734–8), vol. 1, 3.

[20] Bayle (1734–41), preface. Osborn (1938), 33 says 889 new lives were added, of which Birch wrote 618.

[21] Bayle (1734–41), vol. 1, dedication page; Osborn (1938), 33–4 for list of Birch's entries.

[22] Ibid, vol. 7, 776–802; vol 6, 219.

letters of Newton and John Wallis, and the entry on Henry Oldenburg drew heavily on his correspondence with Boyle. But there is no evidence of a concern with how the person's scientific ideas developed, and certainly no indication of the possibility of situating an individual's work in relation to that of contemporaries, as was suggested in William Wotton's planned, but uncompleted, biography of Boyle.[23] There was no clear acknowledgment of science as a special vocation, distinct from other kinds of scholarship.[24]

Biographical dictionaries: the Biographia Britannica

The major English biographical dictionary of the eighteenth century was the *Biographia Britannica*, published between 1747 and 1766, and again in a second edition between 1778 and 1793. Edward Gibbon paid it a somewhat ambiguous compliment, saying that: 'The author of an important and successful work may hope without presumption that he is not totally indifferent to his numerous readers: my name may hereafter be placed among the thousand of articles of a *Biographia Britannica*.' This was 'the first book in any language having the title of *Biographical Dictionary*'.[25] It excluded the geographical and historical material covered by Moreri and its founding editor, William Oldys, aimed to give 'a more methodical Collection of *Personal History*', 'in the manner of Bayle'.[26] Both editions defended the inclusion of men whose lives revolved around ideas, rather than military and political action. The first edition noted that 'we have very few memorials of PHYSICIANS, though scarcely any nation has produced better', and it advertised the inclusion of the 'most eminent Scholars, with a clear and rational account of their works'. It mentioned 'Men of Letters' as an important category, and 'Philosophers, Physicians, Mathematicians, Chemists etc' were said to be neglected by previous dictionaries of this kind. One practical rationale was supplied: namely, that when men of genius realised that their contributions would 'not be buried in oblivion' they would be 'more eager in pursuit of knowledge and virtue'. Nevertheless, the coverage of men of science in the *Biographia Britannica* looks weak when compared with that of a more

---

[23] Osborn (1938), 26 on the use of letters; Hall (1949) on Wotton; also Hunter, this volume. See Stauffer (1941), 252 on the lack of critical sense in Birch's expansion of Bayle.
[24] See Shapin (1991)
[25] Christie (1884), 202; Stauffer (1941), 249 on the others. For Gibbon's remark, see Reese (1970), 3.
[26] Preface to first edition by Oldys reprinted in Kippis (1778–93), vol. 1, xiv.

specialised work such as Benjamin Martin's, *Biographia Philosophica* of 1764.[27]

The *Biographia Britannica* also gave explicit attention to the scope and role of biography. At least in the preface by Oldys there was an interest in the *development* of a reputation: detailed personal histories of famous individuals, it averred, allow us to trace 'the beginnings of their greatness, and learn the steps by which they rose'. Kippis claimed that 'Biographical knowledge' like 'Natural Philosophy' could move from individual cases to 'general truths and principles'.[28] But this did not translate into a serious analysis of their intellectual activities. Paul Korshin has remarked that 'one can read dozens of lives in the *Biographia Britannica* without uncovering a single notion as to what somebody thought, how he composed, or whom he knew'.[29] While this may be a little extreme in the case of some of the entries on philosophers and men of science, it is generally true that scientific publications were treated as marks of *action* — the category operating for the more numerous entries on political and military figures. The entries on Boyle and Newton are quite detailed; the one on Newton occupies thirty-four pages and includes some discussions of his work, together with a liberal use of letters. This solid article here no doubt reflects the amount of material available for summary, some of it written by Newton's disciples such as Henry Pemberton and Colin Maclaurin.[30] However, most entries on scientific figures – such as those on Barrow, Bentley, Derham, Maclaurin, Ray – mention their scientific works but give no extensive summary of these, of developments in their thought, or of any debates in which they were involved. Instead, the evaluative emphasis centres on their character, with most being seen as examples

---

[27] Preface to first edition reprinted in Kippis (1778–93), vol. 1, xiv–xvi. Stauffer 1941, 257 says Kippis was more receptive to literary figures than Oldys. See Stanley (1701), preface, for the view that the Ancients gave greater attention to men of thought and contemplation. Martin (1764) has entries on Roger Bacon, Francis Bacon, Barrow, Bernoulli, Boyle, Brahe, Copernicus, Roger Cotes, Descartes, Flamsteed, Galileo, David Gregory, James Gregory, Halley, Hobbes, Hooke, Horrox, Huygens, Kepler, Leibniz, Locke, Maclaurin, Newton, Oldenburg, Pascal, Rohault, Torricelli, John Wallice [sic], Whiston, Wilkins, and Wren. Some of these give summaries of the theories associated with each person – especially those of Copernicus, Kepler, Galileo, Descartes, Newton – and their place in the history of science. Martin acknowledged the accounts by Colin Maclaurin and Henry Pemberton. See Pemberton (1728) and Maclaurin (1748); and Stewart, L. (1992), for the context of this popularisation of Newton.

[28] Oldys in Kippis (1778–93), vol. 1, xv. Kippis (1778–93), preface to 2nd edition, vol. 1, xxi. For comments on this work as an example of early literary biography, see Stauffer (1941), 249–56 and Lipking (1970), 79–81.

[29] Korshin (1974), 516.

[30] Oldys (1747–66), vol. 2, 913–34 on Boyle; vol. 5, 3210–44 on Newton. See Pemberton (1728); Maclaurin (1748).

of the 'Christian Philosopher'.[31] Significantly though, in the biographical entries this particular concept does seem to be most closely linked with *natural* philosophy – a nexus which, as Steven Shapin has noted recently, was not so clear in Burnet's original use of it in his sermon at Boyle's funeral.[32]

Nevertheless, the articles in this dictionary were longer and more detailed than those in Moreri or in Birch's ten-volume English edition of Bayle. There was also more sense of the contemporary context in which an individual operated, and Oldys recognised the possibility of 'a succinct account of any disputes or controversies in which they were engaged'.[33] But in Kippis's second edition, in particular, these accounts ran out of control, lost in irrelevant facts pursued under the compulsion for what Boswell called 'authentick information'.[34] A nineteenth-century critic said that Kippis did not rewrite the first edition of the dictionary 'by methodising those lives which were injudiciously or incorrectly given in the first edition', but rather gave the article verbatim and then added his additions and corrections. This gave 'the whole the air of a tedious controversy between himself and the preceding editors.' One consequence was that the second edition only reached volume five, ending with the letter F.[35]

Biography in encyclopaedias

By the late eighteenth century some lines of demarcation between historical dictionaries and encyclopaedias had collapsed. Biography now appeared on both sides of the border. The second edition of the *Encyclopaedia Britannica* departed from the original edition by including biographical articles along with those on the major 'systems' of the arts and sciences. This was sufficient to cause the resignation of William Smellie, the compiler of the first

---

[31] Ibid, vol. 3, 1649; vol 5, 3047, 3494; also Kippis (1778–93), vol. 5, 116 on William Derham. The concept of 'character' did not necessarily prescind such intellectual analysis. For its role in David Hume's historical works, appearing during the period of the first edition of the *Biographia Britannica*, see Wertz (1993).

[32] Shapin (1993), 338–9 and (1994), 170–2. This point would need to be confirmed by a more thorough check of a variety of entries in both Birch's edition of Bayle and the *Biographia Britannica*.

[33] Oldys (1447–66), vol. 1, xiv.

[34] Kippis (1778–93), vol. 1, xix aimed for biographies based on 'the most original information, to render them peculiarly authentic'. On Boswell, see Stauffer (1941), 402–55 and Dowling (1978); Boswell (1953), 22–3, 694, 770.

[35] Chalmers (1812–17), vol. 17, 382–86 at p. 384. The warning signs appeared in volume two, where Kippis had to explain why the letter 'B' had so many names. Kippis (1778–93), vol. 2, vii. Lipking (1970), 85 says that English works of this kind 'never resolved the mixture of antiquarian curiosity and critical ambition on which they were predicated'.

edition – or at least it was the reason he gave for severing his ties with the publication he had joined at the invitation of two Edinburgh printers, Andrew Bell and Colin Macfarquhar. One reporter says that when his partners insisted upon the introduction of 'a system of general biography' in the second edition, Smellie withdrew with the objection that this was 'by no means consistent with the title *Arts* and *Sciences*.'[36]

This incident highlights the assumption about the role of encyclopaedias as compendia of rational systems, contrasting with other dictionaries, such as the historical ones. Thus when biography *was* included in encyclopaedias from the end of the eighteenth century, it marked the introduction of foreign material, information belonging to a previously separate genre of reference publication – namely, the historical dictionary. Biographical articles were a new kind of entry – neither technical terms nor large systems – and thus fitted awkwardly into the dictionaries of arts and sciences that appeared from the mid eighteenth century, either as successors of Harris' *Lexicon* and Chambers' *Cyclopaedia*, or as smaller versions of the *Encyclopédie*.

In the second edition of the *Encyclopaedia Britannica*, edited by James Tytler in ten volumes between 1778 and 1784, the preface announced the appearance of biography. In doing so the editor stressed that it was 'a new department', not found in 'any collection of the same kind'.[37] He gave an enthusiastic endorsement of the genre, noting its sheer human interest and its moral value. The longest biographical entry (five and a half columns) on a scientific figure was, unsurprisingly, the one on Isaac Newton. This gave the standard account of the main moments in his life, referring to his work on astronomy and Biblical chronology, but not the work on optics or alchemy. However, like the earlier dictionaries, it still made 'character' the governing concept: Newton's avoidance of scientific controversy and his modesty about his intellectual achievements were treated under this rubric. Any detailed account of his scientific theories was still reserved, as in the first

---

[36] Kerr (1811), vol. 1, 363. Smellie did however write biographies of contemporaries such as David Hume. These were published posthumously in Smellie 1800. In a letter of 1797 about his own work on Adam Smith, Dugald Stewart said: 'I hate biography'. Stewart (1980), 265. For another criticism of anecdotal biography by Stewart, see Stauffer (1941), 538–9.

[37] Tytler (1778–84), vol. 1, vii. In fact, 'biography' is listed on the title page of a work which appeared in 1774. See Proctor and Castieau (1774). I have not been able to check the extent to which this advertising is confirmed by the contents. Tytler's edition of the *Britannica* does not use this word, but the last part of the extended content description on the title page promised 'an Account of the *Lives* of the most eminent Persons in every Nation, from the earliest ages down to the present times.' As far as I can tell, the word 'biography' never appeared on the title page of any edition of the *Britannica*.

edition, for the entry on 'Newtonian Philosophy' which followed immediately after the biographical one.[38] This separation of the biography of a scientific figure from an account of his scientific work (although less extreme in the case of Newton) set the pattern for the biographical entries in the *Encyclopaedia Britannica* between the 1780s and the 1820s.[39]

The inclusion of biography was justified by a link with the main content of the encyclopaedia – the large 'systems' on the various arts and sciences. Thus the preface suggested that 'after surveying any particular science, it will be found equally useful and entertaining to acquire some notion of the private history of such eminent persons as have either invented, cultivated or improved, the particular art or science'.[40] In giving the policy on the selection implied here, the editors of the third edition (Colin Macfarquhar, and later George Gleig) said that the biographical entries would treat persons who have 'distinguished themselves either in the theatre of action or in the recess of contemplation'. They accepted that this rule might lead readers to complain about the absence of 'their favourite philosopher, hero, or statesman', or conversely, about the inclusion of some obscure names 'who were no proper objects of such public regard'. But they insisted that selection was determined by the presence of some link 'with recent discoveries and public affairs'.[41]

Thus the second and third editions of the *Britannica* – appearing between 1778 and 1797 – did set limits on the scope of biography. The preface to the two volume *Supplement* of the third edition, undertaken by the Episcopal clergyman, George Gleig in 1801 (a second edition was issued in 1802), is quite revealing. Explaining the omission of 'articles (chiefly biographical)' which were advertised in the prospectus, Gleig admitted that he had to deviate from the original plan of supplementing the third edition with sketches of men whose lives did not meet the existing test for inclusion – that is, having some connection with 'science, art, or literature.' He now admitted

---

[38] Ibid., vol. 7, 5385–87 on Newton; 5388–99 on Newtonian philosophy.

[39] The entry on Roger Boscovich in the third edition seems more integrated than most, but recent research suggests that it was written in two parts. Michael Barfoot of Edinburgh University Library has referred me to Goldie (1991) who claims that Patrick Geddes wrote the article. However, Barfoot (private communication) argues that there is a case that Geddes wrote only the biographical part, and that John Robison wrote the pages on the scientific theories. For the Boscovich entry, see Macfarquhar and Gleig (1788–97), vol. 2, 92–9 on his life and character; then 99–107 begins: 'It now remains that one give an account of his Theory of Natural Philosophy'.

[40] Tytler (1778–83) vol. 1, vii. This sentence was repeated in the prospectus to the third edition, issued in 1788. See British Library: Mic. B. 896, reel 1505, no. 24.

[41] Macfarquhar and Gleig (1788–97), vol. 1, x–xi. This third edition was issued in 300 parts from 1788 and was completed as eighteen volumes in 1797.

that his subsequent experience confirmed the appropriateness of the stricter, original criterion:

> So many applications were made to me to insert accounts of persons who, whatever may have been their private virtues, were never heard of in the republic of letters, that I was under the necessity of excluding from the second volume [of the Supplement] the lives of *all* such as had not either been themselves eminent in literature, or in some liberal art or science.[42]

This could be taken as a vindication of Smellie's stand against biography. It certainly showed the danger of a subject that touched human emotions and left editors to deal with the pleas of readers to perpetuate the memory and reputation of their dead friends and relatives. Nevertheless, James Millar, the editor of the fourth edition, which appeared from between 1801 and 1810, seemed to loosen the previous policy. Like his predecessors, he justified the inclusion of some 'obscure' names on the grounds that 'there has rarely passed a life of which a faithful record would not be useful'. But by combining this statement with the announcement that the encyclopaedia now aimed for a 'more perfect biographical register than any which has hitherto been offered to the public', he threatened the existing rationale – the need for some link with a major discovery (in the case of men of science) as the condition for the inclusion of a biographical entry.[43]

It is possible that the introduction of biography allowed the *Britannica* to meet the challenge of its smaller rivals. Apart from Abraham Rees' editions of Chambers' *Cyclopaedia* (which appeared in various editions of four or five folio volumes between 1778 and 1788), it was undoubtedly the major English language encyclopaedia of the late eighteenth century; but it began to face competition from a variety of publications that experimented with size and price in order to attract sections of the market. For example, *The New Royal Encyclopaedia Londonensis*, appearing between 1796 and 1802, claimed to fit the best of the *Britannica* into three volumes – by *excluding* biography. It asserted that 'very expensive Works (stimulated by private Interest instead of public Utility) have thus loaded their Performances, by absurdly . . . running into a tedious Prolixity of Systems of BIOGRAPHY, GEOGRAPHY, ENGLISH, HISTORY etc' which take up three-quarters of the

---

[42] Gleig (1801), vol. 1, vi.
[43] Millar (1801–10), vol. 1, xiii. This was a reprint of the third edition, plus a two volume supplement.

work and are not 'proper to form a Part of an Encyclopaedia of Arts and Sciences'. A similar charge was brought by a larger rival — *The English Encyclopaedia: a collection of treatises and a dictionary of terms, illustrative of the Arts and Sciences*, issued in ten volumes in 1802. In its view, the *Britannica* had 'been swelled with such a variety of uninteresting biography; with tedious geographical descriptions of obscure towns and villages; with minute histories of fabulous heroes and divinities.' On the other hand, the *Encyclopaedia Perthensis, or Universal Dictionary of Knowledge* [1815] included biography because arts and sciences 'are far from comprehending every necessary subject of inquiry'.[44] But if they were to remain smaller and cheaper, most of these competitors could not afford to include biography while also covering the expanding material in the arts and sciences. The *Britannica* was prepared to do both; indeed, by 1802 it had no choice because Rees began his own projected 44 quarto volume *New Cyclopaedia*, thus dramatically breaking the tie with Chambers *and* including biographical articles.

Biography and history of science

Whatever the commercial strategy behind the decision of the *Britannica* to embrace biography, there were significant intellectual consequences. The justification given for biography led to a related case for the historical and geographical contexts of science. Thus the second edition predicted that most readers, having learnt something of the persons involved in a 'branch of human knowledge', will naturally wish to know 'something of the places where those transactions have passed'.[45] The preface of the fourth edition (repeating a sentence from the prospectus to the third edition) asserted that there was a 'natural and necessary connexion' between major achievements and the 'scenes where they were performed'.[46] This association between person, place and intellectual activity was also offered in Richard Pulteney's *Historical and Biographical Sketches of the Progress of Botany in England*, published in 1790. His preface noted that:

> In tracing the progress of human knowledge through its several gradations of improvement, it is scarcely possible for an inquisitive and liberal mind, of congenial taste, not to feel an ardent wish of

[44] [Howard] [1796–1802], vol. 1, vi; anon. (1802), vol. 1, vii; anon. [1815], vol. 1, ii.
[45] Tytler (1778–84), vol. 1, vii.
[46] Millar (1801–10), vol. 1, xiii; see also McKillop (1965) on the notion of local attachment in the eighteenth century.

> information relating to those persons by whom such improvements have severally been given; and hence arises that interesting sympathy which almost inseparably connects biography with the history of each respective branch of knowledge.[47]

This assumption meant that biography could function as a Trojan horse for more historical treatments of science in encyclopaedias.

But did this happen? What form did this treatment of biography take? In what follows I offer some remarks on the place of biographies in encyclopaedias — against the background of the earlier historical and biographical dictionaries. As already noted, the *Encyclopaedia Britannica* began to carry biographical entries after 1778 and these included some on men of science. I have not been able to carry out a comprehensive survey of the proportion of scientific lives in these editions, but I can offer some observations. It is clear that no living figures were included. This followed the accepted view of biographical dictionaries which, as the English edition of Bayle said, were 'a kind of general monument, to the memory of deserving persons'.[48] Biographical entries were final accountings of a life; they were memorials having both a testimonial and moral function. The editors of the third edition put one of the consequences rather starkly, explaining that some important figures were missing because the letters of their names had passed before 'we had intelligence of their deaths'. Similarly, when David Brewster and the Reverend John Lee were compiling the *Edinburgh Encyclopaedia*, Lee offered to write an entry on the Scottish historian and philosopher, Adam Ferguson — on the expectation that he would die before they reached the letter 'F'.[49] Another implication was that during the last two decades of the eighteenth century the *Britannica* carried no separate biographical entries on the leading contemporary figures of science, some of the heroes of what is now regarded as a second scientific revolution.[50]

However, in the large treatises or 'systems' on particular sciences, the contributions of key contemporary figures *were* mentioned, although still not

[47] Pulteney (1790), vol. 1, cited in Stauffer (1941), 505–6.
[48] Bayle (1734–8), vol. 1, dedication; Kippis (1778), vol 1, xviii. But presumably editors had to consider the 1721 House of Lords decision that it was a breach of privilege to print the 'life of a peer without the consent of his heirs or executors.' Plant (1965), 120.
[49] Macfarquhar and Gleig (1788–97), vol. 1, x. Lee to Brewster, 6 December 1809, MS 3432, ff 231–2, National Library of Scotland.
[50] See Kuhn (1961), 190 Cohen (1985), 97 on the idea of a second scientific revolution. Writing in 1845 Henry Brougham claimed that scientists of this period surpassed those of seventeenth century. Brougham (1855–61), vol. 1, v.

as part of a biographical treatment. For example, the article on 'Chemistry' in the third edition of the *Britannica* opens with an historical introduction organised around the contributions of those, beginning with Boyle, who freed the subject from its alchemical shackles.[51] There is a separate biographical entry on Boyle, but more recent participants, such as Joseph Priestley, Joseph Black, William Cullen, Henry Cavendish and Antoine Lavoisier, are not covered in biographical entries, although their theories and disputes are mentioned in 'Chemistry' and in some of the other articles, such as 'Aerology'.[52] With the appearance of the *Supplement* to this third edition in 1801, a biographical notice of Lavoisier (who died on the guillotine in May 1794) was included, and more space was given to his and other recent work in the large article on chemistry.[53]

When he began to compile the *Supplement* to the fourth, fifth and sixth editions of the *Encyclopaedia Britannica* in 1815, Macvey Napier introduced new biographical articles, 'mostly of recent lives'. Although some additions filled what he called 'palpable omissions' in the previous editions, Napier focused mainly on 'persons who have died during the last thirty years', and in doing so brought the number of biographies to 165 within a total of 600 articles in the six volumes of the *Supplement* completed in 1824. Of these biographical entries, about a quarter were the new ones Napier added: that is, on people who had recently died. He explained that 'the subjects of them [biographical articles] have been selected, for the most part, on account of their eminence in Science or Literature'.[54] In fact, of the 165 lives 58 (35%) were men of science. In contrast, only 5 (3%) of the biographical articles were on religious figures.[55] This seems to indicate Napier's preferences, although generalisations have to be tempered by the fact that 55% of all biographical articles in the *Supplement* were drawn from the first three letters of the alphabet. Another contingency is that forty-six of all the biographies

---

[51] Macfarquhar and Gleig (1788–97), vol. 4, 374–635; see 374–7 for mention of Boyle.

[52] Ibid., vol. 4, 394; vol. 1, 144–97. In the article on chemistry, Black, Cullen, Cavendish, Kirwan, Boerhaave are mentioned in both the index and the text.

[53] Gleig (1801), vol. 1, 210–403; this article takes up half of part one of volume 1. See also vol. 2, 70–2 for a very positive article on Lavoisier. There was no biography here of Priestley, Black or James Watt.

[54] Napier (1824), vol. 1, xxix.

[55] Gooding (1929), 66–7. In the case of the *Edinburgh Encyclopaedia*, started in 1809 but not completed until 1830, Brewster's coeditor, John Lee, drew up an alphabetical notebook (dated 1805) with forty-seven possible entries, mainly biographical ones, finishing at the letter T. These seem to be suggestions for lives not adequately covered in other encyclopaedias, but only two, Barrow and Robison, are natural philosophers. MS 3455, National Library of Scotland. By the time it appeared Brewster had remedied this to some extent.

were written by Thomas Young, the natural philosopher and Egyptologist. But there is little doubt that Napier was concerned to enlarge the representation of scientific lives in the encyclopaedia.[56]

Charles Babbage referred in 1830 to the 'diminutive' world of science.[57] This observation holds, even allowing for the polemical intent of commentators such as Babbage who wrote about the alleged decline of science in England in comparison with France. In the early nineteenth century, the number of individuals who could be counted as men of science was small when compared with other vocational categories. It follows that Napier must have been well disposed towards scientific lives for the biographical entries: otherwise the figure of 35% could not have been achieved. He also went beyond England and Scotland. Thus it is worth noting that the *Encyclopaedia Britannica*, unlike the *Biographia Britannica*, was not a strictly British reference work; indeed, its coverage of leading French and German figures in the biographical entries was quite extensive. This was complemented by the fact that Napier recruited foreign experts as contributors to some of the major articles on the sciences: Francois Arago on 'Polarisation', Jean Baptiste Biot on 'Pendulum', 'Electricity' and 'Galvanism'. On both counts the work presented itself as continuing the ideal of a cosmopolitan republic of learning at a time when biographical dictionaries, for both practical and political reasons, had become decidedly national.[58]

What was the nature of these biographical entries in Napier's *Supplement*? How do they compare with those in the 'historical' and biographical dictionaries? Any answers must first recognise that although Napier made a point of adding to the existing biographical articles of the *Britannica*, these were not his highest priority, even if they may have played a part in a marketing strategy. Like most early nineteenth-century encyclopaedia editors, he devoted most of his energy to the task of recruiting appropriate contributors for the major treatises on scientific and other subjects. Once engaged, some of these authors volunteered biographical pieces on men of science associated with the field they were covering: for example, James

---

[56] Napier (1824), vol. 1, xxix for mention of Young as responsible for 'a large proportion of such as relate to men of science'. See Peacock (1855), vol. 2, 446–623 for a selection of Young's biographical contributions.

[57] Babbage (1830), x–xi.

[58] Of the 165 biographical articles in Napier's *Supplement* at least 94 were on Continental European figures. For the earlier cosmopolitan ideal, see Daston (1991). The *Encyclopaedia Metropolitana*, projected by S. T. Coleridge in 1817, at first intended to arrange its historical sections in chronological order, starting with the ancient world; but the editors soon shifted to national histories. See Smedley *et al.* (1845), vol. 1, xviii.

Smith did some of those related to botany. Napier said that some accounts of distinguished figures were offered by friends who had 'long personal knowledge of them'. One effect of this pragmatic approach was that the biographical entries of the previous editions were republished, usually without revision, whereas alterations and additions to those on the various arts and sciences were a selling point of successive editions.[59] This meant that Napier did not recommission the entries on the major scientific personalities of the seventeenth century who were included in earlier editions. Thus the entry on Newton remained unaffected by new archival evidence regarding his alchemical views and heterodox theology until Brewster wrote for the seventh edition of the *Britannica*, completed in 1842.[60] Given these factors, it is unlikely that a concerted approach to biography can be discerned, but there are features that distinguish encyclopaedias from the earlier historical and biographical dictionaries.

The *Britannica* never extended its initial short notice of the word 'biography' into a separate article until the eleventh edition of 1910. Napier did contemplate this when preparing his *Supplement*, and asked William Hazlitt: 'Is there anything *good* and *striking* to be said upon *Biography*, its kinds, rules and cases?'[61] But this article never materialised. Nevertheless, the entry for James Boswell served as a discourse on the positive features of the genre. 'We commemorate him as an author, and particularly as a writer of Biography. Here he is almost an inventor'.[62] But the 'minuteness' of information, mentioned as one of the key features of this modern biographical approach, and attempted by Kippis in the *Biographia Britannica*, was rarely sought in the biographical entries of the *Supplement*. The extensive footnotes and quotations from letters were also omitted. One practical reason here was the limited space available in the *Supplement*: Napier cited this when asking Hazlitt to 'curtail' the length of a biographical entry.[63] Yet within the space allotted, choices of emphasis were made. The biographies of scientific figures ranged in length from under a page to five pages (making the latter ones

---

[59] Napier (1824), vol. 1, xxx; Gooding (1929); Yeo (1991), 47.

[60] See Brewster (1842). These issues were more fully confronted in his biography of Newton. See Brewster (1855) and Christie (1984). Augustus De Morgan also wrote an essay on Newton for a collection called *Old England's Worthies*. This was informed by his own extensive knowledge of the available archives. The material was seen as controversial, so De Morgan signed his name to it. De Morgan (1846). For the contemporary debate about Newton's character, see Yeo (1988) and (1993), ch. 5.

[61] Napier to Hazlitt, 24 August 1816, MS 674, f 69, National Library of Scotland.

[62] Napier (1824), vol. 2, 372–75, p. 373.

[63] Napier to Hazlitt, 24 August 1816, MS 674, f. 69, National Library of Scotland.

substantial entries of about 3000 words), but the focus was always on their scientific work rather than on their personal lives.

This suggests a contrast with the historical and biographical dictionaries in which the character of the person was usually more prominent than the details of their scientific or scholarly work. We should note, however, that the eighteenth-century notion of 'character' was closer to our idea of reputation and was not synonymous with later concepts of personality as the expression of inner psychological dynamics, often developing from childhood. Further consideration of this issue is beyond my scope here; but it is possible to say that character was often regarded as something directly displayed.[64] In many of these eighteenth-century publications, biographical articles on philosophers invariably referred to the physical body of the person: that is, to stature, health, diet, habits. Thus according to Moreri, Aristotle ate little 'and slept less'; Bacon's 'Port was Stately, his Speech flowing and grave'; Boyle 'was of a weak infirm Body, which renders it the more astonishing how he could write, meditate, read, and try Experiments as he did'.[65] The English edition of Bayle continued this concern: Galileo was 'of little stature but of venerable aspect and vigorous constitution. His conversation was affable, and free, and full of pleasantry.' On the other hand, Robert Hooke's person 'was but despicable, being very crooked, and always pale and meagre'.[66] Similarly, moral character was deduced from behaviour. These entries also allowed space for character assessments by contemporaries, including accounts of the last days of the person and their deathbed demeanour. In Kippis, this emphasis on character, both physical and moral, continued, but with greater attention to the doctrines and controversies associated with the person, often illustrated with long extracts from their letters. By the third edition of the *Encyclopaedia Britannica* (from 1788), this emphasis on character had not altered, although it offered much shorter biographical articles than the major eighteenth-century historical and biographical dictionaries.

---

[64] For some comments on the concept of 'character' as used by Hume, see Wertz (1993). D'Alembert's definition in the *Encyclopédie* points towards the modern sense: 'In a moral sense it signifies an habitual disposition of the soul, that inclines to do one thing preferably to another. Thus a man who *seldom* or *never* pardons an injury, is of a revengeful character. Observe, we say seldom or never, because a character results not from a disposition being rigorously constant at all times, but from its being generally habitual, and that by which the soul is the most frequently swayed.' D'Alembert (1772), 301. On nineteenth-century notions, see Collini (1991), ch. 3.

[65] Moreri (1694), no pagination.

[66] Bayle (1734–41), vol. 5, 372; vol. 6, 218. This portrait of Hooke was repeated in Oldys (1741), vol 4, 2659.

By the time of Napier's *Supplement* (1815–24) however, even the intention of discussing character seems to have gone. Thus the entry on Charles Coulomb states that: 'Mr Coulomb's moral character is said to have been as correct as his mathematical investigations', but there was no attempt to discuss it; indeed, this statement followed five pages summarising nineteen of his scientific memoirs and papers.[67] A passage from the entry on the eighteenth-century Russian natural philosopher Francis Aepinus confirms the priorities at work here:

> We regret that our means of information do not enable us to communicate any particulars in regard to his personal history; but we shall give some account of his contributions to science, and these, after all, form the most interesting memorials of a philosopher's life.[68]

In Napier's *Supplement*, the emphasis was on the contribution to the story of scientific progress; men of science were considered as *discoverers*, not as individuals whose whole lives were being weighed. The concern with details about the body and demeanour of the person was now usually absent; so too were the long extracts from their letters or character assessments by friends. As Peter Carey noticed in the extract from his novel quoted at the top of this chapter, there was rarely any room for the idiosyncrasies of the particular man of science. The articles in the *Supplement* to the *Britannica* began to distance 'the merely personal' – as Albert Einstein later called it – from its account of scientific lives. How far this set the direction for the treatment of scientific biography in other encyclopaedias, and indeed, later editions of the *Britannica*, remains a question for investigation. For example, some comparative analysis of Abraham Rees *New Cyclopaedia*, which included biographical entries, would be useful. Charles Hutton's *A Philosophical and Mathematical Dictionary* (1815) also advertised 'Memoirs of the Lives and Writings of the Most Eminent Authors, both ancient and modern, who by their discoveries and improvements have contributed to the advancement of them'.[69]

One reason for the limited detail about personal lives in the case of

---

[67] Napier (1824), vol. 3, 414–19.
[68] Ibid, vol. 1, 63.
[69] See Bernstein (1985), 296. For a comment on the evacuation of self from nineteenth-century scientific biography, see Gagnier (1991), 258. For the late eighteenth-century contrast, see Outram (1978). As might be expected, biographies in the Positivist *Calendar* of Auguste Comte evaluated the individual's contribution to the progress of science and gave little space to their views on other matters. See Harrison (1892).

Napier's work is the way he severely restricted the theological glosses that accompanied character portraits of natural philosophers in earlier publications such as the English edition of Bayle and the *Biographia Britannica*. (I have already cited the low percentage of religious leaders and theologians.) In his *Supplement*, biographical accounts of men of science very rarely include references to the natural theological framework that support the praise of their behaviour as Christian philosophers. This contrasted with the practice of the earlier English biographical dictionaries and with the evangelical emphasis in contemporary early nineteenth-century British society. Indeed, at some points Napier's *Supplement* seemed closer to the anti-clerical outlook of the French *Encyclopédie* – for example, the negative effects of fanatical religion were mentioned in some articles. Thus in noting that the Italian mathematician and philosopher, Maria Agnesi, retreated to a convent towards the end of her life, the writer cited this as 'another melancholy instance' of the 'darkening power of superstition over the brightest minds'.[70]

But the main reason for this reduced interest in personality was the new function of these short biographical entries as satellites to the larger articles on the various sciences.[71] Under this arrangement, biography was subservient to the history of science. As mentioned above, a large part of each entry under the name of a natural philosopher recorded his or her publications; it also ranked the person against other individuals using the more detailed account of the progress of science given in the larger articles on particular disciplines. Thus Coulomb was evaluated 'in the particular department of science which he cultivated', and it was concluded that 'he may be fairly ranked in the same class with Franklin, Aepinus, and Cavendish'.[72] The grounds for this assessment were to be found in the article on 'Electricity'. In retrospect, we can see a hint of this division of territory between biographical and scientific articles in the third edition of the *Britannica*. For example, the biographical entry on Copernicus pointed to the one on Astronomy for a summary of his theory, but there was no assessment, or even discussion, of the scientific work of a person in the short biographical notices.[73] In Napier's *Supplement* however, many of these entries awarded their subject, at least in a shorthand way, a position in the history of the relevant field of science – as

---

[70] Napier (1824), vol. 1, 113. See Carew (1831) for a response to Stewart's comments on rational religion, and Hilton (1988) for evangelical attitudes.

[71] For an analysis of the organisation of encyclopaedias, see Yeo (1991).

[72] Napier (1824), vol. 3, 419.

[73] Macfarquhar and Gleig (1788–97), vol. 5, 432–33 referring to vol 2, 421. This is only partly qualified in the case of the entry on Newton in the second edition referred to earlier.

this was presented in one of the larger articles, or in one of the three Preliminary Dissertations.[74]

These large 'Discourses on the History of the Sciences', as Napier called them, aimed to provide a 'connected view of the Progress of the Sciences', allowing contrasts and comparisons to be made across some of the disciplines that lay separated by different alphabetical letters in the body of the work. They were not intended to overlook the role of individuals — indeed Napier thought they could highlight the heroic discoverers who received no special place in Chambers' *Cyclopaedia*, those 'great lights of the world by whom the torch of science has been successively seized and transmitted'.[75] But this approach still reinforced the emphasis on the progress of the particular science, not the intellectual or emotional or social biographies of the men of science. Thus in William Brande's dissertation on the history of chemistry 'well known names' appeared alongside the major discoveries he judged as crucial to the progress of the subject; others who failed to grasp, or resisted, this direction had 'sunk into oblivion' and their names were not to be rescued. This approach not only excluded personal details but also various applications of science, if they were not central to a key discovery, as Brande explained:

> I have diligently endeavoured to record every important event in the general history of the science. Of many who have attained deserved eminence in the exclusive pursuit of its distinct branches, no mention has been made: I have looked with attention into their works, and am well aware of their individual merits; but I should have swerved from the principal object of this Dissertation, *that of recording discoveries*, had I attempted even the superficial enumeration of their infinitely varied applications.[76]

Conclusion

When biography entered the *Encyclopaedia Britannica*, in the second edition from 1778, this was recognized as a departure from the prevailing convention that allocated this subject, as well as more general history and geography, to the 'historical' dictionaries, such as those following Moreri's. There

---

[74] On these see Yeo (1991), 33–4 and (1993), 150–1.
[75] Napier (1824), vol. 1, xxxi.
[76] Brande (1818), vol. 3, 50, 79; my emphasis.

was also the suggestion that readers would appreciate information on the historical and geographical settings in which major personalities acted.[77] However, this did not take the form of a life history as found in the earlier historical or biographical dictionaries, and by the time of Napier's *Supplement* (1824) the biographical articles largely abandoned any attention to the character or personality of the individuals they noted. The focus was on their contributions to the march of discoveries surveyed in the preliminary dissertations and explained in the 'treatises' on the various sciences.

It would be pointless to assess this approach to biography in terms of modern conceptions of the subject. However, we can try to view the encyclopaedic treatment of biography in the context of early nineteenth-century notions about what an account of scientific lives might contain. In 1845 Henry Brougham gave the following prescription in his *Lives of Men of Letters and Science, who flourished in the time of George III*:

> The history of a philosopher's life, that is, of his labours, the tracing of those steps by which he advanced beyond his predecessors, the comparison of the state of the science as he found it, with that in which he left it, tends mightily to interest the reader, to draw him towards the same inquiries.[78]

These requirements were to some extent met by the biographies in Napier's *Supplement* to the *Encylopaedia Britannica*. Especially when taken in tandem with the larger articles on the sciences and the preliminary dissertations, Brougham's concern with ranking an individual discoverer is apparent in Napier's work. Although such accounts are now seen as overly 'presentist' or Whiggish, they were attuned to the 'state of science' at different historical periods, and hence fulfilled a need recognized by some of the leading members of the scientific community. Commenting on discussions at the British Association over the history of chemistry in 1845, George Peacock stressed the importance of explaining why a theory [phlogiston] now known to be false 'was so long considered to be true'; this involved, he said, an awareness of the precise conditions of opinion prevailing in specific periods of scientific history.[79] On a related issue, William Vernon Harcourt, the first secretary of the British Association, criticised Brougham's handling of the controversy over the roles of Black, Watt and Cavendish in determining the chemical

---

[77] See note 43 above and Macfarquhar and Gleig (1788–97), vol. 1, x–xi.
[78] Brougham (1855–61), vol. 1, ix–x.
[79] Peacock (1845–6), 108, 132.

composition of water. Writing to Brougham he suggested that such issues were complex and that if venturing on 'such dangerous ground, you should at least learn how to choose your authorities'.[80] The articles in the 1824 *Supplement* to the *Encyclopaedia Britannica* did, to some extent, provide the starting point for such historical inquiries relevant to what Harcourt called 'arbitration of the rights of discovery'.

From this perspective biographies did not need to ponder the character of the man of science, his social or religious views, nor even his reasons for such an unusual vocational choice. The 'lives' in encyclopaedias were small sections of the history of science in which the contributions of an individual were mapped against a larger story. There is a paradox here in so far as this historical genealogy (or even 'rational reconstruction') of discoveries coexisted with an increasing interest in the heroic genius of great discoverers.[81] Thus in early Victorian biographical writing outside encyclopaedias there was concern with the question of intellectual style and its connection with personal and private life. This became apparent when separate books (as distinct from dictionary entries or obituaries) began to be written on men of science. In 1831 Humphry Davy's biographer, John Paris, argued that an understanding of scientific figures would not be found in a collection of anecdotes, but 'in an analysis of human genius, and in the development of those elements of the mind, to whose varied combinations, and nicely adjusted proportions, the mental habits, and intellectual peculiarities of distinguished men may be readily referred'.[82] Another example from the middle of the century suggests that the celebration of the intellectual qualities of major scientific individuals still left an image of the man of science as remote from life, absorbed in strange journals and language, 'as one who refuses to conform to the conventionalities of society, rejects its enticements'. As an antidote to this perception, a writer for the *Eclectic Review* recommended Francois Arago's biographies of six French scientists (Baily, Fourier, Carnot, Malus, Fresnel, Laplace) who, 'while they pursued the most occult subjects of scientific research, were, for good or evil, foremost in the political movements of their age'.[83] A study of the biographical essays on major scientific

---

[80] Harcourt to Brougham, in Harcourt (1846). For a critical review on more general grounds, see Lockhart (1845) who said the 'public might reasonably expect more care' from Brougham about the history of science.

[81] Schaffer (1986).

[82] Paris (1831), vol. 1, iv, vii; also Knight (1992) and his chapter in this volume. On the emergence of popular scientific biographies, see Sheets-Pyenson 1990, 399.

[83] Anon. (1857), 201–2.

figures in the Victorian periodicals, and in scientific journals such as the *Philosophical Magazine*, might provide instructive comparisons with those in encyclopaedias.

These examples illustrate the variety of opinion on the appropriate form of scientific biography. Even as he announced his own view, Brougham's conception of this genre as a way of allotting the individual a place in the history of discovery did not satisfy all readers or authors. They had to look beyond encyclopaedias for a different approach, but also beyond the earlier 'historical' and biographical dictionaries which focused on moral character in the absence of any detailed account of the scientific work of the person. Similarly, as historians or biographers, we must choose our eighteenth- and nineteenth-century evidence carefully. Perhaps one of the lessons from this study is the need to appreciate the rationale of dictionaries and encyclopaedias before using them as primary sources for biographical material, or as evidence of contemporary images of scientists.

Acknowledgements

This study was supported by an Australian Research Council grant, the Ian Potter Foundation, and by the Faculty of Humanities, Griffith University. For research assistance I thank Judith Deppeler-Hagan and Diana Solano; for research involving French translation I thank Alice Addison. I am also grateful to John Gascoigne, Dorinda Outram, Michael Shortland and Bill Zachs for their comments.

Bibliography

Note: Articles in nineteenth-century journals appeared anonymously and have been identified in Houghton (1966–88).

Aiken, J. and Enfield, W. (1799–1815) *General Biography; or, Lives, Critical and Historical*, London: G. and J. Robinson.
Alsted, J. (1989) [1630] *Encyclopaedia . . . Serie Praeceptorum, Regularum, et Commentarium Perpetua*, ed. with foreword by W. Schmidt-Beggemann. Stuttgart–Bad Cannstatt.
Anon. (1802) *The English Encyclopaedia: being a Collection of Treatises and a Dictionary of Terms, Illustrative of the Arts and Sciences*, 10 vols. London: G. Kearsley.
Anon. [1815] *Encyclopaedia Perthensis; or, Universal Dictionary of Knowledge, collected from every source*, 23 vols. London: C. Mitchell.
Anon. (1857) Biographies of distinguished scientific men. *Eclectic Review*, 2, 201–25.

Arago, F. (1857) *Biographies of Distinguished Scientific Men*, trans. W.H. Smyth, B. Powell and Robert Grant. London: Longman.

Babbage, C. (1830), *Reflections on the Decline of Science in England*. London: B. Fellowes.

Bacon, F. (1974) [1605] *The Advancement of Learning and the New Atlantis*, ed. A. Johnson. Oxford: Clarendon Press.

Bayle, P. (1710) *An Historical and Critical Dictionary, by Monsieur Bayle. Translated into English with many Additions and Corrections, made by the Author himself, that are not in the French Editions*, 4 vols. London: C. Harper and others.

Bayle, P. (1720) [1697] *Dictionnaire Historique et Critique*, 3rd edition, 4 vols. Rotterdam: Chez M. Bohm.

Bayle, P (1734–8) *The Dictionary Historical and Critical of Mr Peter Bayle. The Second Edition*, trans. P. Des Maizeaux, 5 vols. London: D. Midwinter.

Bayle, P. (1734–41) *A General Dictionary, Historical and Critical*, ed. J.P. Bernard, T. Birch, J. Lockman, and other hands, 10 vols, London: J. Bettenham for G. Strahan.

Bayle, P. (1826) *An Historical and Critical Dictionary, selected and abridged from the Work of Peter Bayle: with a Life of Bayle*, 4 vols. London: Hunt and Clark.

Bayle, P (1965) *Historical and Critical Dictionary: selections*, trans. with introduction by R. Popkin, New York: Library of Liberal Arts.

Bernstein, J. (1985) The merely personal. *American Scholar* **54**, 295–302.

Boswell, J. (1953) *Life of Johnson*. Oxford: Oxford University Press.

Bradshaw, L.E. (1981) Ephraim Chambers' Cyclopaedia. In Kafker (1981), pp. 123–40.

Brande, W.T. (1818) The preliminary dissertation on the progress of chemical philosophy from the early ages to the end of the eighteenth century. In Napier, vol. 3, pp. 1–79.

[Brewster, D.] (1842) Newton. In *Encyclopaedia Britannica*, 7th edition, 21 vols. Edinburgh: A. and C. Black, vol. 16, pp. 175–81.

Brewster, D. (1855) *Memoirs of the Life, Writings and Discoveries of Sir Isaac Newton*, 2 vols, Edinburgh: Constable.

Brougham, H. (1845–6) *Lives of Men of Letters and Science, who Flourished in the Time of George III*, 2 vols. London: C. Knight.

Brougham, H. (1855–61) *Works*, 11 vols. London and Glasgow: R. Griffin.

Burrell, P. (1981) Pierre Bayle's *Dictionnaire Historique et Critique*. In Kafker (1981), pp. 83–103.

Carew, Rev. P.J. (1831) *Remarks, Analytical and Historical, on the Connexion of Revealed Religion with Literary and Civil Liberty; being a Reply to Certain Statements in the First Preliminary Dissertation of the Encyclopaedia Britannica*. Dublin: Richard Coyne.

Carey, P. (1988) *Oscar and Lucinda*, Brisbane: University of Queensland Press.

Chalmers, A. (1812–17) *The General Biographical Dictionary, Containing an Historical and Critical Account of the Lives and Writings of the Most Eminent Persons in every Nation*, 32 vols. London: J. Nichol and Son *et al.*

Chambers, E. (1728) *Cyclopaedia, or An Universal Dictionary of Arts and Sciences*, 2 vols. London: J. Knapton *et al.*

Christie, R. (1902) *Selected Essays and Papers*. London and New York: Longman, Green and Co. Reprinted from Biographical Dictionaries. *Quarterly Review*, 157 (1884), 187–230.

Christie, J.J.R. (1984) Sir David Brewster as an historian of science. In A.D. Morrison-Low and J.J.R. Christie (eds) *Martyr of Science: Sir David Brewster 1781–1868*. Edinburgh: Royal Scottish Museum, pp. 53–6.

Cohen, I.B. (1985) *Revolution in Science*. Cambridge, MA: Harvard University Press.

Collier, J. (1721) *The Great Historical, Geographical, Genealogical and Poetical Dictionary; being a Curious Miscellany of Sacred and Profane History*, 3rd edition. London: H. Rhodes.

Collini, S. (1991) *Public Moralists: Political Thought and Intellectual Life in Britain, 1850–1930*, Oxford: Clarendon Press.

[D'Alembert, J.] (1772) Analysis of the word character. In *Select Essays from the Encyclopedy, being the most curious, entertaining, and instructive parts of that very extensive work, written by Mallet, Diderot, D'Alembert, and others, the most celebrated writers of the age*. London: Samuel Leacroft.

Daston. L. (1991) The ideal and reality of the Republic of Letters in the Enlightenment. *Science in Context*, 4: 367–86.

De Morgan, A. (1846) Newton. In *Old England's Worthies*, London: C. Knight, vol. 11, pp. 78–117. [I have used a proof copy in the De Morgan papers, Senate Libary, London University.]

Diderot, D. and D'Alembert, J. (1751–80) *Encyclopédie, ou Dictionnairé Raisonné des Sciences des Arts et des Métiers*, 35 vols, Stuttgart–Bad Cannstatt.

Dowling, W.C. (1978) *Boswell and the problem of biography*. In D. Aaron (ed.) *Studies in Biography*. Cambridge, MA: Harvard University Press, pp. 73–93.

Elwin, W. (1855) Arago and Brougham on men of science. *Quarterly Review*, 97, 473–513.

Gagnier, R. (1991) *Subjectivities: a History of Self-Representation in Britain 1832–1920*. Oxford: Oxford University Press.

Galloway, T. (1844) The martyrs of science. *Edinburgh Review*, 80, 164–98.

Gleig, G. (ed.) (1801) *Supplement to the Third Edition of Encyclopaedia Britannica*, 2 vols. Edinburgh: T. Bonar.

Goldie, M. (1991) The Scottish Catholic Enlightenment. *Journal of British Studies*, 30, 20–62.

Gooding, L.M. (1929) The *Encyclopaedia Britannica*: a critical and historical study of the three Constable editions. Unpublished Master of Science thesis, Toronto University.

Harcourt, W.V. (1846) *Letter to Henry Brougham FRS containing remarks on certain statements in his Lives of Black, Watt and Cavendish*, London: R. Taylor. Copy in William Whewell Papers, 103. c. 80.1 no. 2, Trinity College, Cambridge.

Hall, A.R. (1949) William Wotton and the history of science. *Archives Internationales d'Histoire des Sciences*, 9, 1047–62.

Hall, W. (1797) *The New [Royal] Encyclopaedia; or Modern Universal Dictionary of Arts and Sciences, on a New Improved Plan*, 3 vols, 3rd edition, revised by T.A. Lloyd. London: C. Cooke.

Harrison, F. (ed.) (1892) *The New Calendar of Great Men. Biographies of the 558 Worthies of all Ages and Nations in the Positivist Calendar of Auguste Comte*. London: Macmillan.

Hilton, B. (1988) *The Age of Atonement: the Influence of Evangelicalism on Social and Economic Thought 1785–1865*. Oxford: Clarendon Press.

Houghton, W. (1966–88) *The Wellesley Index to Victorian Periodicals*, 5 vols. Toronto: Toronto University Press.

[Howard, G.S.] [1796–1802], *The New Royal Encyclopaedia Londinensis; or, a Complete Modern and Universal Dictionary of Arts and Sciences*, 3 vols. London: printed for A. Hogg.

Hunter, M. (1990) Alchemy, magic and moralism in the thought of Robert Boyle. *British Journal for the History of Science*, **23**, 387–410.

Hutton, C. (1815) *A Philosophical and Mathematical Dictionary*, 2nd edition, 2 vols. London.

Jeffrey, F. (1848) The discoverer of the composition of water: Watt or Cavendish?. *Edinburgh Review*, **87**, 67–137.

Kafker, F. (ed.) (1981) *Notable Encyclopedias of the Seventeenth and Eighteenth Centuries: Nine Predecessors of the Encyclopédie*. Oxford: the Voltaire Foundation, *Studies on Voltaire*, vol. 194.

Kafker, F. (1994) William Smellie's edition of the *Encyclopaedia Britannica*. In F. Kafker (ed.) *Notable Encyclopedias of the Late Eighteenth Century: Eleven Successors of the Encyclopédie*. Oxford: the Voltaire Foundation, *Studies on Voltaire*, vol. 315, pp. 145–82.

Kerr, R. (1811) *Memoirs of the Life, Writings and Correspondence of William Smellie*, 2 vols. Edinburgh: J. Anderson.

Kippis, A. (1778–93) *Biographia Britannica; or, the Lives of the Most Eminent Persons who have flourished in Great Britain and Ireland*, 2nd edition, 5 vols. London: W. and A. Strahan.

Knight, D.M. (1992) *Humphry Davy: Science and Power*. Oxford: Blackwell.

Korshin, P.J. (1974) The development of intellectual biography in the eighteenth century. *Journal of English and Germanic Philology*, **73**, 513–23.

Kuhn, T.S. (1961) The function of measurement in modern physical science. *Isis*, **52**, 161–93.

Laertius, D. (1954) *Lives of Eminent Philosophers*, with a translation by R.D. Hicks. Cambridge, MA: Heinemann.

Lipking, L. (1970) *The Ordering of the Arts in Eighteenth-Century England*. New Jersey: Princeton University Press.

Lockhart, J.G. (1845) Lord Brougham's lives of men of letters. *Quarterly Review*, **76**, 62–98.

Macfarquhar, C. and Gleig, G. (1788–97) *Encyclopaedia Britannica; or, A Dictionary of the Arts and Sciences, and Miscellaneous Literature*, 3rd edition. 18 vols. Edinburgh: Bell and Macfarquhar.

Maclaurin, C. (1748) *An Account of Sir Isaac Newton's Philosophical Discoveries, in four books*. London: printed for the author's children by A. Millar.

Martin, B. (1764) *Biographia Philosophica, being an Account of the Lives, Writings, and Inventions of the Most Eminent Philosphers and Mathematicians who have flourished from the Earliest Ages of the World to the Present Time*, London: W. Owen.

Millar, J. (1801–10), *Encyclopaedia Britannica; or, a Dictionary of Arts, Sciences, and Miscellaneous Literature, enlarged and improved*, 4th edition, 20 vols. Edinburgh: A. Bell for A. Constable and Company.

Miller, A. (1981) Louis Moreri's *Grand Dictionnaire Historique*. In Kafker (1981), pp. 13–52.

McKillop, I.D. (1965) Local attachment and cosmopolitanism: the eighteenth-century pattern. In F.W. Hilles and H. Bloom (eds) *From Sensibility to Romanticism: Essays Presented to F.A. Pottle*. New York: Oxford University Press, pp. 191–218.

Moreri, L. (1694) *The Great Historical, Geographical and Poetical Dictionary; being A Curious Miscellany of Sacred and Profane History. Containing, in short, the Lives and most*

Remarkable Actions . . . of ancient and modern authors; of Philosophers, Inventors of Arts, and all those who have recommended themselves to the world, by their Valour, Virtue, Learning, or some Notable Circumstances of their Lives, 2 vols. London: printed for Henry Rhodes.

Napier, M. (ed.) (1815–24) *Supplement to the Fourth, Fifth, and Sixth Editions of the Encyclopaedia Britannica*, 6 vols. Edinburgh: A. Constable.

Oldys, W. (1747–66) *Biographia Britannica; or, the Lives of the Most Eminent Persons who have flourished in Great Britain and Ireland*, 6 vols. London: W. Innys.

Osborn, J.M. (1938) Thomas Birch and the *General Dictionary* (1734–41). Modern Philology, 36, 25–46.

Outram, D. (1978) The language of natural power; the *éloges* of Georges Cuvier and the public language of nineteenth-century science. *History of Science*, 16, 153–78.

Paris, J.A. (1831) *The Life of Sir Humphry Davy*, 2 vols. London: Colburn and Bentley.

Peacock, G. (ed.) (1855) *Miscellaneous Works of Thomas Young, D.D., F.R.S.*, 2 vols. London: John Murray.

Peacock, G. (1845–6) Arago and Brougham on Black, Cavendish, Priestley and Watt. *Quarterly Review*, 77, 105–39.

Pemberton, H. (1728) *View of Sir Isaac Newton's Philosophy*. London: S. Palmer.

Plant, M. (1965) *The English Book Trade: an Economic History of the Making and Sale of Books*, 2nd edition. London: George Allen and Unwin.

Priestley, J. (1765) *A Description of a Chart of Biography; with a catalogue of names inserted in it, and the dates annexed to them*. Warrington.

Proctor, P. and Castieau, W. (1774) *The Modern Dictionary of Arts and Sciences; or, Complete System of Literature*, London.

Pulteney, R. (1790) *Historical and Biographical Sketches of the Progress of Botany in England from its Origin to the Introduction of the Linnaen System*, 2 vols. London.

Reed, J.W. (1966) *English Biography in the Early Nineteenth Century 1801–1838*. New Haven: Yale University Press.

Rees, A. (1802–20) *The New Cyclopaedia; or, Universal Dictionary of Arts and Sciences*, 44 vols. London: Longman.

Reese, M.M. (1970) *Gibbon's Autobiography*. London: Routledge.

Retat, P. (1991) Encyclopédies et dictionnaires historiques au XVIII siècle. In A. Becq (ed.) *L'Encyclopédisme: Actes du Colloque de Caen, 12–16 Janvier 1987*. Paris: Editions Aux Amateurs de Livres, pp. 505–11.

Rose, H.J. (ed.) (1840–8) *A New General Biographical Dictionary*. London.

Schaffer, S. (1986) Scientific discoveries and the end of natural philosophy. *Social Studies of Science* 16, 387–420.

Scott, J., Green, C. and Meader, J. (1765) *A General Dictionary of Arts and Sciences; or a Complete System of Literature*. London: S. Crowder.

Shapin, S. (1991) 'A Scholar and a Gentleman': the problematic identity of the scientific practitioner in early modern England. *History of Science*, 29, 279–327.

Shapin, S. (1993) Personal development and intellectual biography: the case of Robert Boyle. *British Journal for the History of Science*, 26, 335–45.

Shapin, S. (1994) *A Social History of Truth: Civility and Science in Seventeenth-Century England*. Chicago: University of Chicago Press.

Sheets-Pyenson, S. (1990) New directions for scientific biography: the case of Sir Willian Dawson. *History of Science*, 27, 399–410.

Smedley, E., Rose, Hugh. J. and Rose, Henry J. (eds) (1829–45) *Encyclopaedia Metropolitana; or, Universal Dictionary of Knowledge, on an Original Plan*, 29 vols. London: Fellows and Rivington.

[Smellie, W.] (1768–71) *Encyclopaedia Britannica; or, A Dictionary of the Arts and Sciences, compiled upon a New Plan*, 3 vols. Edinburgh: Bell and Macfarquhar.

Stanley, Thomas (1701) *The History of Philosophy. Containing the Lives, Opinions and Actions, and Discourses of the Philosophers of Every Sect*, 3rd edition, London: W. Battersby.

Stauffer, D.A. (1930) *English Biography before 1700*. Cambridge: Cambridge University Press.

Stauffer, D.A. (1941) *The Art of Biography in Eighteenth-Century England*. New Jersey: Princeton University Press.

Stewart, D. (1980) [1796] *Account of the Life and Writings of Adam Smith* (ed. by I.S. Ross). In W.P.D. Wightman and J.C. Bryce (eds.), *Adam Smith: Essays on Philosophical Subjects*. Oxford: Clarendon Press, pp. 265–352.

Stewart. L (1992) *The Rise of Public Science: Rhetoric, Technology and Natural Philosophy in Newtonian Britain, 1660–1750*, Cambridge: Cambridge University Press.

Strauss, G. (1966) A sixteenth-century encyclopaedia: Sebastian Munster's *Cosmography* and its editions. In C. Carter (ed.), *From the Renaissance to the Counter-Reformation*. London: J. Cape, pp. 145–63.

Tytler, J. (1778–84) *Encyclopaedia Britannica; or, A Dictionary of Arts, Sciences, etc. On a Plan Entirely New* 2nd edition, 10 vols. Edinburgh: Bell and Macfarquhar.

Wertz, S.K. (1993) Hume and the historiography of science. *Journal of the History of Ideas*, **54**, 411–36.

Whiteley, S. (1992) The circle of learning: *Encyclopaedia Britannica*. In J. Rettic (ed.), *Distinguished Classics of Reference Publishing*. Phoenix, Arizona: Onyx Press, pp. 71–88.

Yeo, R. (1988) Genius, method and morality: images of Newton in Britain, 1760–1860. *Science in Context*, **2**, 257–84.

Yeo, R. (1991) Reading encyclopaedias: science and the organisation of knowledge in British dictionaries of arts and sciences, 1730–1850. *Isis*, **82**, 24–49.

Yeo, R. (1993) *Defining Science: William Whewell, Natural Knowledge and Public Debate in Early Victorian Britain*. Cambridge: Cambridge University Press.

Zedler, J.H. (ed.) (1732–50) *Grosses Vollstandiges Universal-Lexikon*, 64 vols. Leipzig and Halle.

6

The scientist as hero: public images of Michael Faraday

GEOFFREY CANTOR

Like many readers of this chapter my early encounters with science were not only through the school laboratory and classroom but also through reading biographies of scientists. Excepting the adventures of Biggles and Gimlet, my serious reading as a child was largely confined to biographies of Newton, Davy, Pasteur and many others, often in the form of collected biographies, such as Egon Larsen's *Men who Changed the World* (1954) and J.G. Crowther's *Six Great Scientists* (1955).[1] Such works kindled my interest in science and played some role (which I leave to my future biographers to specify) in my choice of career, first as a physicist and subsequently as an historian of science. Yet as a young consumer of biographies I would have had no appreciation of the complexities of biographical narratives or of their cultural, educational and ideological functions. I simply lapped them up and contented myself by reading about the Great and the Good and how they contributed to science.

More recently I have re-engaged the subject of scientific biography not only as a consumer but also as a producer, in writing a biographical study of Michael Faraday, in which I paid particular attention to the interrelation between his science and his religion.[2] In pursuing that project I became aware of a vast and untapped treasury of material that fell outside the usual range of sources discussed by historians of science, or indeed biographers. I started to read some of the numerous obituary notices that followed Faraday's death in 1867 and the biographies and biographical sketches that have been published, especially those aimed at the wider reading public, including

[1] Larsen (1954); Crowther (1955).
[2] Cantor (1991).

those written for children — works with such enticing titles as *Heroes of Science*, *Pioneers of Electricity* and *Men who Found Out*.[3] Too often such sources are simply ignored by professional historians of science since they were rarely written by famous authors and usually shed no new light on Faraday's life and works. Yet the apparent uniformity of these biographical works should not mask their historical significance, for they indicate how Faraday was perceived by his contemporaries and successors.

Even more importantly, they show how 'Faraday' was *constructed* in the public arena. I place Faraday in quotation marks since I am not concerned with the historical Faraday who was born on 22 September 1791 but rather with a number of different 'Faradays' who were the purported subjects of these biographical narratives, nearly all of which were written after his death on 25 August 1867. Closer inspection of this literature shows that these 'Faradays' fulfilled many different and even contradictory functions. Thus, for some authors, he became the great discoverer of nature's secrets, while for others he was the Christian philosopher *par excellence*, or the leading public lecturer, or the scientist with refined sensibilities — to mention but a few. These portrayals of Faraday — or more exactly these 'Faradays' — embody complex cultural values and meanings. They posit the nature of science, its aims and methods, and also the ideals for which the scientist should strive. Faraday thus usually emerges as the exemplary scientist, not only in the methods he employed in analysing nature but also as an individual who exemplified the highest personal, social and religious values. One subtext of such biographical narratives is that readers — especially prospective scientists — should adopt Faraday's methods and attitudes as their model. As one writer stated, 'For a student of science, or anyone having a mild interest in natural philosophy, the life of a man like Faraday is an inspiration . . . [His life] should be widely known and emulated'.[4]

Yet I also want to emphasise that despite the manifest similarities between these popular biographies they differ substantially, not so much in the detailed information they contain, but in the way their authors interpret Faraday. This short chapter cannot do justice to such diversity; I will accordingly confine discussion to two opposing types of narrative. One I shall call 'Romantic', the other 'Realistic' — both terms being used in a broad sense, except when more closely specified. (A significant proportion of texts falls

---

[3] Garnett (*c.* 1885); Munro (1890); Williams-Ellis (1929).
[4] Anon. (1901).

clearly into one or other genre, although some biographies and biographical sketches of Faraday combined both modes vicariously.) These two genres provide different accounts not only of Faraday but also of the nature of science and the personality of the scientist.[5] Each therefore conveys a different message to the reader. In the ensuing discussion I will pay particular attention to the different constructions of Faraday's early life and those factors identified as responsible for his success as a scientist.

Tyndall's romance

John Tyndall's *Faraday as a Discoverer*, one of a clutch of books and articles that followed shortly after Faraday's death, deserves particular attention since it exemplifies the Romantic construction of Faraday and was extensively cited by later writers in this genre. One reviewer noted, Tyndall's 'story . . . has all the charms of a romance', while another claimed that he portrayed Faraday as 'a pre-Raphaelitic character'.[6] Tyndall was widely read in both German and English Romantic literature. His mentor, Thomas Carlyle, was one of the main channels through which the British reading public – Tyndall included[7] – was introduced to German literature, especially the works of Goethe. Tyndall's attraction to German literature in general, and the writings of Goethe in particular, sprang largely from his concern with the undesirable social and moral effects of utilitarianism and philosophical materialism. He was deeply affected by the riot of hungry mill operatives in Preston in 1842 and by the way their protest was cruelly and bloodily crushed by the army.[8] German idealist philosophy, which Tyndall struggled to understand, appeared to offer an attractive alternative to the socially-destructive philosophies prevalent in Britain.[9] The connection with Goethe is made explicit in the rousing final paragraphs of *Faraday as a Discoverer* where Tyndall states that 'It was my wish and aspiration to play the part of Schiller to this Goethe'. For Tyndall the relationship between these two

---

[5] In this short chapter I am not able to engage the complex problem of locating the Romantic and Realistic accounts of Faraday in their broader cultural contexts. While there has been a long-running dialectic between these two opposing views of the scientist (and of science), they exhibited specific forms in mid- to late-Victorian culture. See, for example, Turner (1974), Hilton (1988) and the papers in the volume edited by Paradis and Postlewait (1985).

[6] Anon (1868*a*, 1868*b*).

[7] Tyndall (1892*d*).

[8] Tyndall (1892*e*). See also Cosslett (1981) and Barton (1987) for useful discussions of Tyndall's Romanticism.

[9] Ashton (1980); Carlyle (1827).

German writers closely parallelled that between Faraday and himself.[10] Moreover, he proceeded to note the similarities between the two men, arguing that, like Goethe, Faraday 'was at times so strong and joyful – his body so active, and his intellect so clear'.[11]

But Goethe, or more accurately the image of Goethe, fulfilled a further desideratum for Tyndall, which can also be traced through Carlyle's influence. Carlyle was obsessed by the notion of the hero. In his *On Heroes, Hero-Worship and the Heroic in History* (1840) he characterised several types of hero, one being the 'Hero as man of letters'. His foremost example was Goethe: 'I consider him to be a true Hero . . . a great heroic ancient man, speaking and keeping silence as an ancient Hero, in the guise of a most modern, high-bred, high-cultivated Man of Letters!'.[12] The absence of scientists from Carlyle's litany of heroes comes as no surprise since he had attacked science as mechanistic and imperialistic in his widely discussed essay analysing the 'Signs of the times' (1829). However, under the influence of Goethe, Tyndall and other scientists Carlyle subsequently came to appreciate that his earlier assessment of science was incorrect and that not all science was mechanistic.[13] While in many of his essays Tyndall likewise interpreted science from a non-mechanistic, Romantic perspective, he also cast Faraday in the mould of a Carlylean hero in *Faraday as a Discoverer*. Tyndall's 'Faraday' was a man of prodigious moral, emotional and intellectual strength: he was a 'genius', 'fiery and strong', no ascetic but animated by an 'excitable and fiery nature'.[14] As well as possessing extraordinary intellectual qualities Faraday was endowed with noble qualities to the highest degree: in his human relationships he was kind and generous, his character was great and chivalrous, and 'his soul was above all littleness and proof to all egotism'.[15] Although he could be hard-hitting when provoked, 'Faraday could not write otherwise than as a gentleman'.[16] Most importantly, he chose the high ideals of science and turned his back on the vast wealth he could have accumulated had he chosen to apply his science for commercial ends.[17] On Tyndall's account he was the exemplary *pure scientist*.

---

[10] Tyndall was not alone in noting the similarity between Faraday and Goethe. John Scoffern (1867, 425–6), a medical chemist, drew attention to the way Faraday, like Goethe, conceived the world to be in constant motion – 'unendliche Bewegung'.

[11] Tyndall (1894), 197.

[12] Carlyle (1891), 146.

[13] Carlyle (1829); Moore (1976).

[14] Tyndall (1894), 172, 90, 176, 43.

[15] Ibid., 122.

[16] Ibid., 43.

[17] Ibid., 181.

Although courageous, the Carlylean hero is not a man of this world but one who transcends it. While Faraday considered himself to be a natural philosopher, Tyndall claimed that he was 'more than a philosopher; he was a prophet, and often wrought by an inspiration to be understood by sympathy alone'.[18] The prophet was included in Carlyle's taxonomy of heroes; unlike ordinary mortals he is 'God-inspired'.[19] Extraordinary qualities feature prominently in Tyndall's description of Faraday; for example, he possessed a 'divining power' that enabled him to look deep into the structure of nature. Such 'flashes of wondrous insight and utterances . . . seem less the product of reasoning than of revelation', while his thoughts about the magnetisation of light were 'more inspired than logical'.[20] These quotations reflect a passage from Sartor Resartus, cited by Tyndall on a subsequent occasion: 'not our Logical, Mensurative faculty, but our Imaginative one is King over us'.[21] The prophet far transcends the bounds of rationality. Among the other supramundane qualities attributed to Faraday were 'spiritual exaltation', 'prescient wisdom' and his remarkable ability for 'making theoretic divination the stepping stone to his experimental results'.[22] In one of the most visionary passages in the biography, Tyndall portrays Faraday playing 'like a magician with the earth's magnetism . . . sweeping his wand [a conductor] across these lines evokes this new power' of electricity.[23]

Not only does Tyndall portray Faraday as unworldly, but he attributes a child-like innocence to him. Even when sorely tried he could write a letter 'as a little child'. Likewise in his science he could express pure delight, 'like that of a boy', when observing common phenomena, such as soap bubbles.[24] But perhaps the most remarkable epithet appeared in Tyndall's discussion of Faraday's illness of the early 1840s: after noting that his subject could no longer engage in society, Tyndall describes him as 'the great Man-child', as if, in his reduced physical condition, the untainted pure soul of the child had reasserted itself.[25]

But there is a related facet of Faraday the seer. While his imagination and transcendental powers took him beyond the knowledge of mere mortals, his unworldliness sometimes prevented him from communicating with others on matters far from common experience. Tyndall thus complained that some

[18] Ibid., 95.
[19] Carlyle (1891), 39. Cf. Knight (1967).
[20] Tyndall (1894), 86, 93.
[21] Tyndall (1892e), 387; Carlyle (n.d.), 218.
[22] Tyndall (1894), 178, 169, 142–3.
[23] Ibid., 34–5.
[24] Ibid., 45, 130.
[25] Ibid., 91.

of Faraday's memoirs on conduction and induction were 'by no means always clear' and that his ideas on the magnetisation of light 'were peculiar to himself, and untranslatable into the scientific language of the time'.[26] On two occasions Tyndall used the evocative contrast between light and darkness to distinguish the true and communicable parts of Faraday's writings from his more obscure and private speculations.[27]

Like other Romantics, Tyndall was particularly concerned with the role of the imagination, a topic on which he addressed the British Association in 1870, citing Faraday as a prime example. With Faraday, he wrote, the imagination's 'exercise was incessant, preceding, accompanying and guiding all his experiments. His strength and fertility as a discoverer is to be referred in great part to the stimulus of his imagination'.[28] In his biography he likewise dwelt on the extraordinary power of Faraday's imagination, which he considered was responsible for his scientific creativity. Thus Faraday 'rose from the smallest beginnings to the grandest ends' in 'virtue of the expansive power which his vivid imagination conferred on him'. Likewise in his researches on frictional electricity Faraday tried 'by the strong light of his imagination to see the very molecules of his dielectrics'.[29]

An active imagination was essential for a discoverer, but many other attributes were also required. Thus Tyndall dwelt on the difficulty of Faraday's work, his 'exquisite skill' and the marvellous precision of his experimental investigations.[30] As a highly accomplished experimentalist Tyndall fully recognised the preparation and skill required for successful experimental work. While he praised Faraday the experimentalist, Tyndall placed great emphasis on the role of theory in Faraday's researches. Indeed, in a Friday evening discourse at the Royal Institution in 1879 he criticised Positivism and claimed that theories are essential for the successful pursuit of science. 'Faraday', he added, 'lived in this ideal world'.[31] In the biography his most succinct statement on the interplay between theory and experiment is the passage where he portrays Faraday's experiments as 'being suggested and guided by his theoretic conceptions. His mind was [also] full of hopes and hypotheses, but he always brought them to an experimental test.'[32]

---

[26] Ibid., 87, 95.
[27] Ibid., 85–6, 169.
[28] Tyndall (1892*b*), 104.
[29] Tyndall (1894), 127, 84.
[30] Ibid., 96, 116, 102, 126.
[31] Tyndall (1892f), 452.
[32] Tyndall (1894), 114.

Tyndall's Faraday was a discoverer — indeed, he was 'the greatest experimental philosopher the world has ever seen'.[33] However, the author's conception of discovery remained staunchly opposed to Positivist mythology.

In discussing Faraday's more mystical qualities, Tyndall went so far as to use metaphors derived from the occult sciences. Thus in his electrical researches, Faraday's work was 'guided by that instinct towards the mineral lode which was to him a rod of divination'. Likewise he 'brooded' for some time on the subjects of electrochemistry and magnetic fields.[34] However, at several places Tyndall drew on metaphors relating to mountaineering and specifically to the Alps. The significance of such metaphors lies not only in the frequent discussions of Alpine scenery by the Romantics but also in Tyndall's experience as a skilled mountaineer who had conquered several of the highest peaks. In one passage he likened Faraday's discoveries to a range of Alpine mountains and explicitly compared the discovery of the magnetisation of light to 'the Weisshorn among mountains — high, beautiful, and alone'.[35] Faraday's work on electromagnetic induction was likewise described as 'the greatest experimental result ever obtained by an investigator. It is the Mont Blanc [which Tyndall climbed three times on scientific expeditions] of Faraday's own achievements.'[36] As both mountaineer and discoverer Tyndall had first-hand experience but he also perceived the close metaphorical connection between these two domains. The scientist is like a climber rising above the city, the common herd and all the mundane aspects of life. Instead he — mountaineering, like science, has strong masculine connotations — samples the pure, untainted delights of nature and savours the fresh air. Faraday was one of this small elite who transcended the multitude.

Other romantic narratives

While Tyndall moulded Faraday into a Carlylean hero and drew explicitly on the Romantic literary tradition, many other authors, who cannot be included among the high Romantics, emphasised more general Romantic themes and values in their biographical accounts of Faraday. Such writers experienced no difficulty in articulating his transition from poverty and social deprivation

---

[33] Ibid., 172.
[34] Ibid., 53, 78, 191.
[35] Ibid., 170–1. Tyndall had been the first person to climb that impressive Alpine peak. See Clark (1981) and Anon. (1868c).
[36] Tyndall (1894), 156.

to the very apogee of British science. This was a variant on the rags-to-riches fable, the very stuff of so many popular novels. More particularly, in 1813, when Faraday was in his early twenties, he was transported from the bindery where he worked to the Royal Institution. This part of Faraday's career, which attracted much comment, was first publicised in John Paris's *Life of Davy* (1831), which printed a letter by Faraday describing his initial contact with Davy in 1812–13 after attending four of his lectures. Here Faraday voiced his 'desire [at that time] to escape from trade, which I thought vicious and selfish, and to enter into the service of Science, which I imagined made its pursuers amiable and liberal'. His wish was soon granted when he was offered the post of Chemical Assistant at the Royal Institution.[37] As one contemporary noted, this transition 'savours almost of the romantic'.[38] The author of an 1868 review likened Faraday's transformation to the discovery of 'the philosopher's stone, which would convert the leaden duties of "trade" into the golden enjoyments of science'.[39]

Later writers often repeated Paris's account, sometimes giving it a bizarre gloss. To quote a children's history book of the 1950s, 'Something like a fairy-story happened next. Davy was very pleased about the notes [Faraday had taken at Davy's lectures] . . . One night there was a loud knock at Michael's door. "Who can that be?" said Michael to himself. He looked out and saw a carriage waiting outside'.[40] It was not Prince Charming but a servant of Humphry Davy's who whisked Faraday off to the Royal Institution. In Floyd Darrow's *Masters of Science and Invention* Faraday's translation to the Royal Institution was achieved with the aid of a magician, not a fairy: 'As by the waving of a magician's wand, here was the young apprentice assisting the master [Davy] and having at his disposal the best equipped laboratory in Europe'.[41] No mortal could have moved with such rapidity from his uncongenial position in the bindery to the mecca of scientific research.

Many romanticised accounts of Faraday's life make him sound like a character from a fairy tale for children. Indeed one writer in the *British Quarterly Review* — hardly a children's periodical — pursued this very theme. After noting that genius springs from all classes, the author conjured up

[37] Paris (1831), ii, 2.
[38] Anon. (1859).
[39] Anon. (1868d), 438.
[40] Magraw (1957), 31.
[41] Darrow (c. 1924), 74.

a beneficent fairy . . . [who must have] paused at the door of a blacksmith in Newington Butts, on the 24th of September, 1791. Silently she lifted the latch, softly she bent over the bed where the new-born infant slept. Unseen she bestowed upon it the mysterious baptism of genius, and straightway little Faraday was consecrated one of the brilliant brotherhood.[42]

In another article, attributed to the naturalist William Crawford Williamson, several romantic themes were emphasised:

We have hitherto referred to Faraday chiefly as a high-priest of Nature, revealing the hidden forces which are her handmaids, and making them manifest to the world. We have seen him grappling successfully with some of the most recondite problems which can engage the human intellect, and manifesting intellectual powers of a character so lofty and subtle that we may regard them as approaching those of a higher order of beings than human . . . [B]y intuitions almost inspired, his mind appears to track the subtle forces which play around our globe . . . It is this combination of the imaginative power with the severe habits of detailed observation that constitutes the highest philosopher.[43]

A dramatic and mysterious event is also recounted in the same article. At a Friday Evening Discourse at the Royal Institution Faraday displayed samples of several liquefied gases in glass test tubes. Although he warned his audience that the mere heat from his fingers might be sufficient to cause the experiment to explode, he tantalised many of those present by keeping firm hold of these volatile test tubes. 'However', concluded Williamson, 'the master magician retained his control over the spirits which his genius had invoked'.[44]

In the preceding quotation Williamson referred to Faraday's 'genius', a term that frequently featured in accounts of Faraday. The way this term came to be applied to natural philosophers in the late eighteenth century has recently been charted by Simon Schaffer. For the Romantics genius is, of necessity, not prosaic but possesses some extraordinary, even mystical,

---

[42] Anon. (1868*d*), 435. Being unfamiliar with the habits of fairies, I don't know why this fairy arrived two days after Faraday's birth.

[43] [Williamson] (1870), 293–4.

[44] [Williamson] (1870), 274. The author was probably referring to a lecture Faraday delivered on 31 January 1845.

power which enables the natural philosopher to divine nature and discover its secrets.[45] Numerous biographers have called Faraday a genius, although not all of them have used the term in its full, Romantic sense. Thus, while Romantics consider genius to be supernatural, one obituarist claimed that he 'owed his exalted position as a man of science wholly to the innate force of his natural genius'.[46]

By the mid nineteenth century the term 'genius' appears to have lost its earlier associations with political radicalism, but it was still sometimes accorded social overtones. While some writers emphasised that geniuses were to be found in all social classes,[47] one contemporary noted that the genius could not be a man of the upper class, born into a life of luxury and educated at the best schools and universities. Instead, the genius is untainted — socially, politically and educationally. Faraday fitted this bill perfectly — or was moulded to fit it. With some satisfaction the writer noted that Faraday, 'the chief Representative Man of the Natural Philosophers in our day', was born into a working-class household. Among the benefits bestowed by this background was a familiarity with the practical manipulations of nature. However, the true philosopher must also transcend his working-class background by means of his intellect and thus he 'ought to be morally elevated, ought to be amiable, ought to be happy'. This was so with Faraday who was 'neither proud nor ashamed of his birth and rearing'.[48]

As we shall see in the next section, Realistic narratives offered a very different account of Faraday's social background. But it is important to notice here that Romantic accounts of Faraday set the innovative scientist on a pedestal and apart from the ordinary man. They also attributed success in science to an ineffable quality that cannot be learnt in the classroom or from reading science books. While the prospective scientist is being encouraged to emulate Faraday as a person, only the chosen few are touched by the divine spark of genius. As one obituarist stated, 'God alone can give genius'.[49] In this respect the scientist is no different from the poet, as several of Faraday's biographers noted. For example, the author of an article in *Once a Week*

---

[45] Schaffer (1990); Knight (1967).

[46] Anon. (1867–8).

[47] Anon. (1868d), 435.

[48] Scott (1860), 207. This was one of the relatively few biographical sketches of Faraday published during his lifetime. By adopting the title 'Representative man' Scott was explicitly modelling his account on Emerson's biographical essays published in *Representative Men* (1850).

[49] Anon. (1867c), 236.

claimed that 'He has (as every great discoverer must have) the imagination of the poet'.[50]

Romantic accounts of Faraday's discoveries emphasise his great imaginative insights into the unity of nature, rather than the details of his empirical researches. For example, in contrast to William Garnett's prosaic account (discussed below), many Romantic themes are to be found in John Scott Russell's review of Tyndall's biography in *Macmillan's Magazine*. Despite its title — 'Faraday, a Discoverer' — this article did not catalogue Faraday's empirical successes; indeed, electromagnetic rotation, electromagnetic induction and the discovery of benzene receive no mention. Instead, for Russell, a civil engineer and shipbuilder, Faraday had discovered much deeper truths about nature which radically changed humankind's understanding of the world. In particular, he was responsible for bringing 'order, harmony, and obedience to law' to the domain of chemistry. Thus solids, liquids and gases — such as ice, water and steam — were no longer treated as separate substances but as the same substance in different states. Again, Faraday had shown that magnetism was not the prerogative of a few substances but was manifested by all matter. Russell's 'Faraday' rejected caloric and brought heat into line with the other physical forces; he showed that chemical bonds could be sundered by electricity and that atoms are polarised. To the domain of physical forces Faraday brought *order*. For Russell such important insights were 'revelations of truth' resulting from an immense spiritual struggle which exhausted Faraday. Moreover, Russell insisted that these revelations could only have been made by a man whose 'whole life is one act of reverence to the Supreme Being in whose presence he finds himself continually illuminated and strengthened'.[51]

Romantic themes also found expression in portraits of Faraday. The most interesting example is the frontispiece to Henrietta C. Wright's *Stories of the Great Scientists* (1889) (Figure 1) which shows 'Faraday announcing his discovery [of electromagnetic rotations] to his wife on Christmas morning, 1821'.[52] The jars, bellows and alembics in the background and the eerie illumination from the door create a mysterious aura. That the discovery should have been made on Christmas morning — a curious misdating that

---

[50] Ibid., 208.

[51] Russell (1868).

[52] Wright (1889), frontispiece. See the portrait of Oersted in Cunningham and Jardine (1990), 229. Paradoxically, while there were a few Romantic allusions in Wright's text, most of her account of Faraday was emphatically Realist.

Figure 1. A romantic portrayal of Faraday. Frontispiece to Henrietta Wright's Stories of the Great Scientists (1889).

first occurred in Tyndall's biography[53] — adds further to the arcane nature of the discovery. Sarah is shown holding a small volume, presumably a Bible, and the couple are caught in a moment of intense intimacy as they witness the simple and marvellous event on the laboratory bench.

Realistic, or Smilesian, narratives

Faraday received passing mention in Samuel Smiles's celebrated *Self-Help* (1859), that casebook of Victorian success and utilitarian philosophy. His brief entry under the heading 'Men who have come from the ranks' provided Smiles with a paradigm example of how perseverance is rewarded. Faraday, 'the son of a blacksmith, was in early life apprenticed to a bookbinder . . . he now occupies the very first rank as a philosopher'.[54] This gloss to the story of Faraday's persistent climb from near the bottom to the top of the social ladder of British science was not original with Smiles; however, I shall for convenience call this type of narrative 'Smilesian' since the ethos of self-help pervades such accounts. Like the Romantic accounts discussed earlier Smilesian narratives are dominantly heroic; however, Smilesian heroes do not possess the supramundane qualities of imagination or ineffable genius. Instead they succeed because they possess determination to the highest degree. Smilesian biographies are anti-romantic and adamantly realist in their portrayal of the hero and his social milieu. The reader is expected to celebrate our hero's climb up the ladder of success, overcoming all obstacles on the way by dint of his own effort.

Since Faraday came from a poor family of no social standing and ended his career as Britain's most celebrated scientist, his biography could readily be fitted into the Smilesian framework. Not surprisingly, many biographies of Faraday were insistently Smilesian and heroic. As one obituarist noted, 'Faraday was a self-made man in the true acceptation of the term. He did not succeed without help, but he made the friends who helped him, and the most valuable help was that he received from himself'. Another claimed that his life has often 'been held up as an example of what self-help can make a man'.[55] On these accounts no supernatural forces aided him during his ascent.

As with Smiles's own writings, this kind of biography usually had a clear

[53] Tyndall (1894), 14.
[54] Smiles (1905), 12, 149–50.
[55] Anon. (1867*a*), 232; Anon. (1867*b*), 182.

and unambiguous message for the reader: if Faraday could achieve such great success by dint of hard work and the use of his native intelligence and other abilities, then anyone with a similar commitment could likewise achieve great things. After all, the qualities Faraday manifested are possessed by all of us. Faraday's life was frequently used for highly moralistic purposes: the story of Faraday's life, commented Frederick James Faraday (a distant relative), 'is one of that good old-fashioned sort which is meet for young boys' reading; the kind of narrative wise men relate to their sons when they wish to plant within them an ambition to be useful men'.[56] Likewise the chemist William Crookes, writing in *Chemical News*, claimed that 'He is the only man, we say, who has raised himself to the first rank in science in this country, whose every attribute we may fearlessly hold up as a model to our children!'[57] Such moralistic intentions were particularly apparent in the numerous volumes of collected biography aimed at young people. It is also significant that several leading medical journals portrayed Faraday in a particularly realistic and naturalistic manner.[58]

Smilesian narratives not only accentuate self-motivation but also provide a structure to the biography by specifying both its starting point in our hero's social origins and an arching framework linking his background and his later successes. Unlike the Romantic genius, the Smilesian hero does not simply spring from heaven (with or without the aid of a beneficent fairy). Instead, despite the poverty and humble origins of his family, one source of Faraday's success can be traced to his parents. Although poor, they already possessed the seeds of honesty and uprightness. They were, in the words of one biographer, both 'respectable . . . and of industrious and excellent habits'.[59] According to another, his father 'was a God-fearing man and a skilful artizan, who, in spite of ill-health and the hard times, endeavoured to give his family some elementary education'.[60] Likewise, Lady Margaret Moir, speaking before the Women's Electrical Society in 1931, drew attention to the influence of Faraday's mother who, despite 'a very rudimentary education . . . had a great struggle in his boyhood to make ends meet and to bring up a young family . . . [Yet] she must have given her son Michael great sympathy and loving encouragement in his early days of struggle'.[61] Although very little is known

---

[56] F[araday] (1867), 501.
[57] [Crookes] (1867), 110.
[58] For example, Anon. (1867*a*); Anon. (1867*d*); Anon. (1869).
[59] Anon. (1869), 636.
[60] Munro (1890), 187.
[61] Scott (1934), 34.

about Faraday's family background these biographers felt sure that his sterling qualities must have been inherited from his parents.

Not only did this generational continuity exist for writers of Smilesian narratives, but they traced Faraday's early life in detail showing how he rose above his origins. These biographers emphasised his lowly social background and portrayed his early life in a realistic, matter-of-fact manner. For example, it was often stressed that the family was forced to accept public relief when the price of corn rose in 1801 and that 'Michael's portion was a loaf, which had to serve him for a week'.[62] Moreover, one children's book claimed that his

> early years were spent in the manner usual to the poorer class of
> city children. He played in the streets with the children of other
> mechanics, and took care of his younger sister while his mother
> was busy about household matters, and ran on errands to neigh-
> bouring shops; and, in fact, had his life filled with that mixture of
> responsibility and duty which usually falls to the lot of the younger
> members of the families of city workmen.[63]

This may be very good character training but there is no historical basis for most of these claims.

Little is known about Faraday's schooling but his Victorian biographers made much of the snatches of information that shed light on his duties as an errand boy and as a journeyman bookbinder. Frequently these writers stressed that Faraday read the books he bound — most famously Jane Marcet's *Conversations on Chemistry* and the *Encyclopaedia Britannica*, which provided his first knowledge of electricity. These introductions to the world of books opened his eyes to science and Faraday made good use of the opportunities available to him in the bindery. Likewise his first attempts at conducting simple electrical experiments using an old glass bottle were frequently presented in a prosaic manner and were used to show that he made the best use of common items available to anyone. Some obituarists, such as Crookes, also dwelt on Faraday's considerable manual skills, learnt while working in the bindery and possibly inherited from his father.[64]

While these Smilesian biographers stressed that Faraday's early experiences were crucially important in shaping the scientist, they also stressed

---

[62] Ibid., 187.
[63] Wright (1889), 226–7.
[64] [Crookes] (1867), 110; Anon. (1867–8), 201.

the positive qualities he possessed which enabled him to rise above his lowly background. Thus Ivor Hart marvelled that despite his lowly birth 'Michael Faraday should have made himself what he did'.[65] How he *made himself* into one of the leading scientists of his generation was a major theme for biographies in this tradition, and they found the key to his success in the admirable, if ordinary, qualities he manifested. One writer claimed that he possessed, to a high degree, such solid qualities as 'good health, an honest purpose, and [an] excellent home training'.[66] Other biographers attributed his success to such qualities as 'his high natural endowments', 'his own energy, [and] perseverance' and his 'persevering industry and spirit of inductive inquiry'.[67] In tune with the ideology of self-help Faraday was seen as a man who had succeeded by hard work, using resources available to even the poorest person. He has often been cited as a paradigm example supporting that ideology. It should come as no surprise that Margaret Thatcher, recently Prime Minister of Britain, hailed Faraday as her hero and kept a bust of Faraday in 10 Downing Street, her official residence.[68]

As noted earlier the meaning of the term 'genius' was not, by the 1860s, confined to its Romantic connotations. In contrast to the gloss given by romantics, Smilesian realists considered genius to be a natural power of the intellect but raised to a high degree. Thus while such writers applied the term to Faraday, they did so in ways that were devoid of transcendental implications. Yet some authors even went so far as to deny explicitly the operation of genius, in the romantic sense of the term, or any other non-rational influence. For example, a writer in the *British Medical Journal* argued that Faraday 'was no juvenile prodigy; he has given proof neither of genius nor even of uncommon talent; but he has shown that he possessed zeal, intelligence, energy, and great steadiness of character . . . [and he] possessed the excellent quality of a slow and steady power of growth'.[69] One rather scathing obituarist in the *Journal of Science* even praised Faraday's strength as an empirical researcher but criticised his more speculative papers. These may have displayed his 'remarkable genius', claimed the writer, but his genius only created useless castles in the air: 'His early education', or want of it, 'unfitted him for large generalisation'.[70]

[65] Hart (1924), 248.
[66] Wright (1889), 227.
[67] F[araday], (1867), 501; Anon. (1867d), 282.
[68] 'Favourite things', broadcast on BBC 2, 26 July 1987.
[69] Anon. (1869), 636.
[70] Anon. (1868e), 54.

Questions of education pervaded many realist constructs of Faraday's life, for not only did they emphasise the steps in Faraday's self-education, but, rather paradoxically, they utilised his biography to teach the reader some elementary science. Such accounts shade into the familiar strategy by science textbook writers of introducing a smattering of biography or history of science.[71] One example of a biography with a definite didactic purpose is William Garnett's *Heroes of Science* (c. 1885) which offers an unenlightening and highly factual account of Faraday's innovations (and contrasts strikingly with John Scott Russell's article, analysed above). Thus we are told that while repeating the experiments described in his 'Historical sketch of electro-magnetism' in 1821 'he discovered the rotation of a wire conveying an electric current around the pole of a magnet'. This innovation is presented by Garnett, who was Principal of the Durham College of Science, as a prosaic extension of Oersted's work which showed that magnets have a tendency to revolve around a current-carrying wire. Adopting an impersonal voice, Garnett continued, 'The principle that action and reaction are equal and opposite indicates that . . . there must [therefore] be an equal tendency for the conductor to rotate around the pole. It was this rotation that constituted Faraday's first great discovery in electro-dynamics.'[72] Garnett's narrative is peppered with terms indicating closure, as if his account were complete and raised no problems for either Faraday or the reader. Thus he claimed that Faraday offered a 'complete explanation' of Arago's wheel, 'demonstrated' electromagnetic induction, and 'saw clearly' that the current induced into a wire depends on the rate at which the lines of magnetic force are cut by the wire.[73] By contrast Garnett expressed little interest in Faraday's theories which 'were based on slender premises, and sometimes were little else than flights of a scientific imagination'![74] Yet he grudgingly accepted that these speculations might have guided his research, but emphasised that they were only significant when verified by experiment. Towards the end of his chapter on Faraday, Garnett noted that his subject tended to separate his religious beliefs from his science. This cursory reference to religion is unexpectedly perfunctory and dismissive considering that Garnett's book was published by the Society for Promoting Christian Knowledge.

Smilesian and realistic themes are also apparent in several illustrations of

---

[71] Kuhn (1962), 1.
[72] Garnett (c. 1885), 255–6.
[73] Ibid., 260–4.
[74] Ibid., 259.

Faraday, such as Figure 2, which appeared in Robert Routledge's *A Popular History of Science* (1881). The central portrait, which is based on the same original as the English £20 note, is a clear, incisive engraving with Faraday staring at the reader. Here is a strong, honest man, dedicated to truth. Routledge's text, studded with pictures of technological devices, is highly factual and avoids all Romantic allusions. Instead it stresses the familiar theme of self-improvement; for example, we learn that Faraday read the books he bound and joined 'a Mutual Improvement Society'. Figure 2 also shows young Faraday reading in the bindery and performing an early electrical experiment. The lower part of the engraving shows Faraday in later life lecturing before a stiff and uncomfortable audience at the Royal Institution. Both of the upper sketches feature books, doubtless important to Routledge, whose family published many cheap novels and utilitarian texts aimed at the mass market.[75]

Epilogue

There was general agreement among writers of both Romantic and Realistic narratives that Faraday could have earned considerable sums of money, but instead nobly preferred science to private gain. As one writer stated, he had 'to decide, whether he should make wealth or science the pursuit of his life. It was a second choice of Hercules. He could not serve both masters . . . He chose the latter and died a poor man.'[76] While his choice of lux over lucre could be incorporated into either narrative form, the subject of utility could not. Romantics portrayed Faraday's science as dedicated only to truth and therefore transcending any practical applications. However, some realists have woven into Faraday's biography the subsequent history of electrical (and sometimes chemical) technology.

This trend has become particularly noticeable this century, especially with the 1931 celebrations marking the centenary of Faraday's discovery of electromagnetic induction. On that occasion many public buildings were floodlit, the British Government threw a large banquet, the Institution of Electrical Engineers held a ball at the Dorchester, celebrities delivered lectures and, most prominently, a massive exhibition was staged at the Royal Albert Hall. But this celebration of Faraday was also a well-orchestrated

---

[75] Routledge (1881), frontispiece & 573–81. See also Barnes and Barnes (1991).
[76] [Dickens?] (1868), 401.

Figure 2. Hard-core realism. Frontispiece to Robert Routledge, A Popular History of
Science (1881).

event by the electrical industry seeking to court the public and expand into the domestic market. The Faraday Centenary proved massively successful and over 100 000 copies of the booklet *Faraday – the Story of an Errand Boy who Changed the World* were sold. However, through these celebrations the public came to know Faraday as the man who had founded the new era of electricity.

Utility has likewise become the dominant theme in many modern popular treatments of Faraday, including some staged in 1991 to coincide with the bicentenary of his birth. For example, the Science Museum mounted an exhibition entitled 'Michael Faraday and the modern world' and a book of the same title was published to coincide with the exhibition.[77] The accompanying poster (Figure 3) shows Faraday surrounded by an electric light bulb (positioned above his head to signify illumination), printed circuits, a compact disc player, television, vacuum cleaner, and other paraphernalia of modern living. The text below this dissonant collage describes Faraday as 'The father of electricity . . . the man who changed the way we live'. This example illustrates a largely twentieth-century twist to Faraday's biography. His life is now rarely valued for either its Romantic connotations or as a paradigm example of self-help. Instead, his discoveries – and their putative impact on technology – have received far greater emphasis than his life and his biography has thereby often been constructed in the service of technology and the powerful electrical industry. Is there not a danger that this kind of writing, which decentres the individual, may destroy the integrity of scientific biography?

Acknowledgements

Early versions of this chapter were read at the Babbage–Faraday Bicentenary symposium in Cambridge (1991), the History of Science Society meeting in Madison, Wisconsin (1991) and the Göttingen History of Science Seminar (1992). My thanks to all those who contributed comments, especially William Clark, Lorraine Daston, Michael Hagner and Skuli Sigurdsson. To Barbara Cantor, Frank James, Michael Shortland, Jonathan Topham and Richard Yeo I am indebted for much information and many helpful comments.

[77] Bowers (1991).

Figure 3. Raising the curtain on Faraday's junkyard. Poster advertising the exhibition at the Science Museum, 1991.

Bibliography

Anderson, P.J. and Rose, J. (eds) (1991) *Dictionary of Literary Biography, vol. 106: British Literary Publishing Houses, 1820–1880*. Detroit and London: Gale Research Inc.

Anon. (1859) Professor Faraday. *Leisure Hour*, 8, 792–5.

Anon. (1867*a*) Professor Michael Faraday. *Medical Times and Gazette*, 18, 232–3.

Anon. (1867*b*) Professor Faraday. *The Engineer*, 24, 182.

Anon. (1867*c*) Professor Faraday. *London Review*, 15, 236–7.

Anon. (1867*d*) Michael Faraday, D.C.L., F.R.S. *Lancet*, 2, 281–2.

Anon. (1867–8) Michael Faraday. *Pharmaceutical Journal and Transactions*, 9, 201–5.

Anon. (1868*a*) Professor Faraday. *London Review*, 16, 289–90.

Anon. (1868*b*) Faraday as a discoverer. *Popular Science Review*, 7, 183–4.

Anon. (1868*c*) Faraday. *All the Year Round*, 19, 399–402.

Anon. (1868*d*) *British Quarterly Review*, 47, 434–74.

Anon. (1868*e*) Faraday. *Journal of Science*, 5, 50–5.

Anon. (1869) Davy's greatest discovery. *British Medical Journal*, 25, 636–7.

Anon. (1901) Review. *Knowledge*, 24, 110.

Ashton, R. (1980) *The German idea. Four English Writers and the Reception of German Thought, 1800–1860*. Cambridge: Cambridge University Press.

Barnes, J.J. and Barnes, P.P. (1991) George Routledge & Sons. In Anderson and Rose, (1991) 261–70.

Barton, R. (1987) John Tyndall, pantheist. A rereading of the Belfast Address. *Osiris*, 3, 111–34.

Bowers, B. (1991) *Michael Faraday and the Modern World*. Wendens Ambo: EPA Press.

Brock, W.H., McMillan, N.D. & Mollan, R.C. (eds.) (1981) *John Tyndall. Essays on a Natural Philosopher*. Dublin: Royal Dublin Society.

Cantor, G.N. (1991) *Michael Faraday: Sandemanian and Scientist. A Study of Science and Religion in the Nineteenth Century*. Basingstoke: Macmillan and New York: St Martin's Press.

Carlyle, T. (1827) State of German literature. *Edinburgh Review*, 46, 404–51.

Carlyle, T. (1829) Signs of the times. *Edinburgh Review*, 49, 439–59.

Carlyle, T. (1891) *On Heroes, Hero-Worship and the Heroic in History*. London: Chapman & Hall.

Carlyle, T. (n.d.) *Sartor Resartus*. London: Harrap.

Clark, R.W. (1981) Tyndall as mountaineer. In Brock *et al.*, (1981), 61–8.

Clubbe, J. (ed.) (1976) *Carlyle and his Contemporaries. Essays in Honor of Charles Richard Sanders*. Durham, NC: Duke University Press.

Cosslett, A.T. (1981) Science and value: the writings of John Tyndall. In Brock *et al.*, (1981), 181–92.

[Crookes], W. (1867) Faraday. *Chemical News*, 16, 110–1.

Crowther, J.G. (1955) *Six Great Scientists*. London: Hamish Hamilton.

Cunningham, A. and Jardine, N. (eds) (1990) *Romanticism and the Sciences*. Cambridge: Cambridge University Press.

Darrow, F.L. (*c.* 1924) *Masters of Science and Invention*. London: Chapman & Hall.

[Dickens, C?] (1868) Faraday. *All the Year Round*, 19, 399–402.

F[araday], F.J. (1867) A great philosopher. *St James's Magazine*, 20, 501–8.

Garnett, W. (*c.* 1885) *Heroes of Science. Physicists*. London: Society for Promoting Christian Knowledge.

Hart, I.B. (1924) *Makers of Science. Mathematics, Physics, Astronomy*. London: Oxford University Press.

Hilton, B. (1988) *The Age of Atonement. The Influence of Evangelicalism on Social and Economic Thought, 1795–1865*. Oxford: Clarendon Press.

Knight, D.M. (1967) Scientist as sage. *Studies in Romanticism*, 6, 65–88.

Kuhn, T.S. (1962) *The Structure of Scientific Revolutions*. Chicago: University of Chicago Press.

Larsen, E. (1954) *Men who Changed the World*. London: Phoenix House.

Magraw, B.I. (1957) *The Thrill of History in Story, Play and Picture. Book IV: 19th and 20th Centuries*. London & Glasgow: Collins.

Moore, C. (1976) Carlyle and Goethe as scientist. In Clubbe (1976), 21–34.

Munro, J. (1890) *Pioneers of Electricity or Short Lives of the Great Electricians*. London: Religious Tract Society.

Paradis, J. and Postlewait, T. (eds.) (1985) *Victorian Science and Victorian Values: Literary Perspectives*. New Brunswick: Rutgers University Press.

Paris, J.A. (1831) *The Life of Sir Humphry Davy, Bart.*, 2 vols. London: Colburn & Bentley.

Routledge, R (1881) *A Popular History of Science*. London: Routledge.

Russell, J.S. (1868) Faraday, a discoverer. *Macmillan's Magazine*, 18, 184–91.

Schaffer, S. (1990) Genius in romantic natural philosophy. In Cunningham and Jardine (1990), 82–98.

Scoffern, J. (1867) Michael Faraday. *Belgravia*, 3, 421–8.

Scott, I (1860) Representative men. The natural philosopher. Faraday. *Once a Week*, 3, 205–10.

Scott, P. (1934) *An Electrical Adventure*. London: Electrical Association for Women.

Smiles, S. (1905) *Self-Help with Illustrations of Conduct and Perseverance*. London: John Murray.

Turner, F.M. (1974) *Between Science and Religion: the Reaction to Scientific Naturalism in Late Victorian Britain*. New Haven and London: Yale University Press.

Tyndall, J. (1892a) *Fragments of Science*, 8th edition, 2 vols. London: Longman, Green & Co.

Tyndall, J. (1892b) Scientific use of the imagination. In Tyndall (1892a), vol. ii, 101–34.

Tyndall, J. (1892c) *New Fragments*. London: Longman, Green & Co.

Tyndall, J. (1892d) Goethe's *Farbenlehre*. In Tyndall (1892c), 47–77.

Tyndall, J. (1892e) Personal recollections of Thomas Carlyle. In Tyndall (1892c), 347–91.

Tyndall, J. (1892f) The electric light. In Tyndall (1892c), 419–52.

Tyndall, J. (1894) *Faraday as a Discoverer*, 5th edition. London: Longman, Green & Co.

Williams-Ellis, A.W. (1929) *Men who Found Out: Stories of Great Scientific Discoveries*. London: Howe.

[Williamson, W.C.] (1870) Review. *London Quarterly Review*, 34, 365–95.

Wright, H.C. (1889) *Stories of the Great Scientists*. London: Ward & Downey.

7

'Tactful organising and executive power': biographies of Florence Nightingale for girls

MARTHA VICINUS

Feminist historians have thoroughly documented the unequal struggle of women who wished to enter the scientific professions, from the very beginnings of modern scientific enquiry in Europe. As long as science remained an amateur occupation of doubtful economic security and without a career structure and public rewards, isolated women from the upper classes — often related to a well-known pioneering scientist — made independent contributions as scientists.[1] But in the nineteenth century matters changed swiftly; while science might still be a hard taskmaster, it brought rewards to many men. Biographies abound of poor boys who found fame as scientists: Michael Faraday, Louis Pasteur, Jean Henri Fabre and others. By the late nineteenth century books aimed at aspiring young scientists appear with titles such as *The Romance of Medicine*, *Microbe Hunters*, *Famous Men of Science* or *Heroes of Modern Progress*. Women were conspicuously absent in these narratives of scientific adventure. But they did not disappear altogether, either in scientific work itself or in the many biographies for girls. These tokens — Caroline Herschel, Mary Somerville, Marie Curie — were eulogised as much for their feminine qualities as their achievements. Ironically, the women who had fought hardest for access to specialised knowledge for the public good — medical women — were rarely mentioned in nineteenth-century biographies.

An examination of the values promoted in these biographies for adolescent girls may help to explain why so few considered a career in science at

---

[1] See especially Schiebinger (1989) for the pre-1800 period and Rossiter (1982) for the United States, as well as the case studies in Abir-Am and Outram (1987) and Glazer and Slater (1987).

the very time when it became possible.[2] The second half of the nineteenth century, when Nightingale hagiography was at its height, saw the growth of an organised women's movement, the development of systematic secondary education for girls, and the opening of institutions of higher education for women. It was also marked by the entry of a few exceptional women into medical and scientific careers. Not only do we have the first women doctors, Elizabeth Blackwell (1821–1910) and Elizabeth Garrett Anderson (1836–1917), but also researchers such as Alice Hamilton (1869–1970), the pioneering American toxicologist, Ida Freund (d. 1914), chemistry teacher at Newnham College, Cambridge and the physicist Hertha Ayrton (1854–1923). But these women rarely if ever appear in biographies for girls, which specialised instead in what were deemed more appropriate examples of female public service. Even though there were real heroines in science, they were not part of the popular imagination, nor did they became role models beyond those they reached personally.

During the second half of the century women reformers actively sought an administrative role in hospitals and philanthropic organisations that was consonant with their domestic experience; they hoped to become rational leaders without losing traditional feminine qualities. Middle and upper class women possessed method, efficiency and order – invaluable skills for improving the civic life of Victorian England. Their undoubted successes stemmed at least in part from contemporary concerns about the haphazard nature of English scientific, medical and military institutions; the shocking death rate in the Crimean war because of organisational failures drew the middle class public's attention to the need for better and more modern administration. In spite of Nightingale's unrelenting efforts to reform the army, change was slow. Male scientists, dismayed at the easy victories of the Prussian army over the French in 1870, called for 'scientific administration' in order to utilise more effectively English talent and material resources.[3] By the end of the nineteenth century, professional men had successfully captured the administrative positions and policy leadership in science and medicine, and were

---

[2] Salmon (1886) lamented the lack of vigorous adventure stories for girls, and strongly recommended biographies: 'Perhaps the best reading which a girl can possibly have is biography, especially female biography, of which many excellent works have been published. One cannot help as one reads the biographies of great women – whether of Miss Florence Nightingale, Mrs Fry, or Lady Russell – being struck by the purity of purpose and God-fearing zeal which moved most of their subjects', 527.

[3] Lockyer (1870). Vicinus and Nergaard (1989), 77–234 document Nightingale's efforts to bring administrative order to the army both in the Crimea and afterwards.

soon to control social work as well.[4] A few women entered the scientific community themselves, but most responded by shifting the terms of debate. If they could not be organisational leaders, they could still control the feminine sphere.

For the Victorians nurses embodied the best elements of the new tenets of scientific organisation and traditional feminine caring. Pioneering nurses became extremely popular figures during the latter half of the nineteenth century. Nursing, an occupation defined for centuries as uniquely feminine, had become decidedly disreputable by mid century; it provided fertile ground for reformers – and their biographers. By picturing untrained hospital nurses as immoral drunkards, often 'the coarsest type of women, not only untrained, but callous in feeling',[5] reformers stressed the dedication of a new generation of women leaders. What could be more attractive than high-minded, well-educated women devoting themselves to raising the moral standards of this 'noble calling'? Florence Nightingale, her protegée Agnes Jones, Sister Dora (Pattison), Dorothea Dix, Clara Barton, and other now-forgotten heroines joined the company of such well-known figures as the famous prison reformer Elizabeth Fry, the Swedish singer Jenny Lind, and the social reformers Agnes Weston, Frances Willard and Catherine Booth. All of the biographies of nurses minimise their scientific ambitions and focus instead on their role as organisers. Nursing, like social work, allowed women a separate sphere of public work that did not impinge upon men's work. Nurses brought an administrative expertise learned in the domestic sphere to the disreputable and inefficient male-dominated hospitals; women alone knew how to combine sympathetic care with organisational skills in order to bring about order and discipline. The actual scientific contribution made by nurses was reduced to their insistence on cleanliness before its benefits were well recognised.

The biographies of Florence Nightingale for girls incorporate the implicit tensions between rational administration and female caring. Those written between the years 1860 and 1914 focus on the adventure of caring for others outside one's immediate family. Those written after World War I provide much more information about her personal struggles and difficulties, rather than her achievements, which are taken for granted. Nightingale embodied characteristics that were especially attractive to the generations before

---

[4] Vicinus (1985), 113–20, 214–18 discusses the gradual loss of female authority and ideological leadership to men in the areas of hospitals and social work.
[5] Quoted in Adams and Foster (1913), 128.

World War I. Her nursing work in the Crimea, occupying only a few months of her long life, seemed to epitomise romantic self-sacrifice for a humanitarian goal; at the same time, she brought the advantages of upper class respectability and powerful political contacts to the reform of this uniquely feminine occupation.[6] Nightingale personified social progress without disturbing social expectations for women. Her intense religiosity, based on social activism, appealed profoundly to a public that believed in doing good for others. She was known to oppose women doctors, as well as numerous new-fangled scientific claims made by medical researchers. Nevertheless, she repeatedly argued for the singular power of a well-trained corps of women nurses. It was a combination that inspired countless popular biographies for girls.

A noble heroine for ambitious girls

By the mid nineteenth century both boys and girls faced vastly increased opportunities. Large numbers of popular biographies poured off the Evangelical presses, offering a variety of role models to emulate, all characterised by a strong religious belief and individual action against odds. Fictionalised anecdotes highlighted the lessons of self-sacrifice and hard work. By providing a series of clear cut moral questions and solutions, the formulaic biographies helped young readers understand the increasingly complex choices they faced. As models for public action, they were an important alternative to the more familiar domestic girls' stories. Indeed, biographies served to validate adolescent dreams of doing good in a wider sphere than the home. Although girls were expected to be more home-loving and nurturing than boys, both sexes had to overcome opposition from either an uncomprehending family or a ridiculing public. The battles were invariably those of an individual pitted against cruel Nature, or bureaucratic obstinacy or an indifferent society. Affectional ties were subordinated to the heroine's public activities. The heroic plot evoked a stereotyped emotional response and provided a vehicle for the reader's personal dreams. The factual narrative of a specific life history combined easily with personal fantasy; the reader did not so much identify with the heroine as with the heroic possibilities she represented.

Although most late nineteenth- and early twentieth-century children's literature had become gender-specific, biographies for girls were often paired

---

[6] These points are made most effectively by Whittaker and Oleson (1978), 19–35 and Poovey (1988), 164–98.

with those for boys; for every *Noble Heroes* collection, there was certain to be *Noble Heroines*. The main figures share many of the same characteristics; courage, pluck, independence, initiative and 'a noble character' are admired, regardless of gender. The sturdy Protestant activism of these stories is revealed in the titles: Joseph Johnson's *Clever Girls of Our Time: and How They Became Famous Women* (1863), *Heroines in Obscurity* (1870) 'by the author of "Papers for Thoughtful Girls" ', *Lives of Good and Great Women* (1888), Edwin Pratt's *Pioneer Women in Victoria's Reign* (1897), Rosa Nouchette Carey's *Twelve Notable Good Women of the Nineteenth Century* (1899), and Jennie Chappell's *Women Who Have Worked and Won* (1904). The same authors — and themes — appear repeatedly; it matters little whether the author was a man or woman. Clearly the writing of such biographies was lucrative hackwork for those who specialised in either religious or children's literature. The emphasis was upon hard work and individual success in this world. These tales demonstrate how a girl through her training and experiences as a child was prepared to face adverse public opinion, misunderstanding and even social ostracism as an adult. The reader was expected to imitate the virtues placed before her, and these were often, as befits a heroine, difficult to master. In the words of one of the most prolific biographers, Joseph Johnson:

> The author's object in writing the following pages has been to incite his sisters — and he includes in that term every member of the gentle sex just merging into womanhood — to imitate or become 'Clever Girls' from the force of example . . . The following pages furnish evidence, if evidence were needed, that woman possesses purpose, will, determination — more than this, that she can attain to a height of perfection in the various professions which enables her to 'hold her own' in the presence of the most gifted of the opposite sex; the 'Clever Girl' merged into the 'Clever Woman'.[7]

Rather than being humbled by circumstances, the heroine learned to control and channel her assertiveness, and thereby control her world.

Nightingale, unlike most heroines, came from a wealthy family; this potential liability was overcome by describing all that she gave up in order to become a nurse. Indeed, her willingness to work for the poor, without losing her class superiority, became one of her chief attributes. The author of

---

[7] Johnson (1863), i, iii–iv.

Women Who Win commented, 'There was every comfort in her home that money could provide, and no one but a genuine philanthropist could have turned away from its attractions for the briefest season. But Providence had other and larger plans for her'.[8] Since self-sacrifice was a key feminine virtue, Nightingale's sacrifice of balls, London night-life and country house parties was seen as a great mark in her favour; the fact that she had loathed these activities was ignored. Nightingale's calculated decision to remain sick after returning from the Crimea in order to be free from family obligations was transformed into the price she had paid for her enormous labours on behalf of the English people. Her lobbying the government to reform the army, public health and sanitation in India was suppressed and all attention focused on her efforts to turn nursing into a respectable female occupation. Such alterations, of course, were common in all popular biographies, which reinforced familiar, stereotyped characteristics in heroines, however far from the truth.

Biographers, anxious to inculcate in their young readers respect for adults, often included an encouraging mentor in the lives of their favourites. For the boy scientist, this most often meant another scientist who could help further his training. Sir Humphry Davy met Michael Faraday when he was still a bookbinder's apprentice; impressed with Faraday's detailed notes of his public lectures, he offered the young man a job.[9] Girls were given moral inspiration, rather than intellectual guidance or economic help. Much was made of Nightingale's brief encounter with Elizabeth Fry, the Quaker prison reformer. As one biographer noted, more honestly than others, 'What passed at that momentous interview we know not, but we may be sure that Florence Nightingale came away stimulated by the elder woman's example and enthusiasm, and encouraged to go steadily forward on the path she had determined to follow for herself'.[10] Sister Dora, the nurse of Walsall, was 'fired with enthusiasm' by 'the story of Florence Nightingale's heroism and self-sacrifice'.[11]

All the heroes and heroines of Victorian children's biographies showed early signs of their vocation, but this was expressed in gendered terms. Darwin was a naturalist by the time he went to school at eight; Jean Henri Fabre filled his pockets with insect specimens from his earliest years.[12] Girls,

---

[8] Thayer (1897), 44.
[9] Bolton (1926), 168–9.
[10] Haydon (1909), 19–20.
[11] Chappell (1898), 95.
[12] Bolton (1926), 233, 289. Obviously, as formulaic biographies, any indecision or difficulties experienced were ignored. Darwin, for example, underwent considerable mental hardship in clarifying

in contrast, eagerly helped the needy. Dora Pattison (Sister Dora) and her sister 'were forever saving their pocket money to give it away, and they made it a rule to mend and remake their old frocks, so as not to have to buy new ones out of their allowance for clothes, so as to have more to give'.[13] One biographer of Nightingale archly described her relationship with her sister, Parthenope: 'The truth is, Florence was born to be a nurse, and a sick doll was dearer to her than a strong and healthy one. So I fear her dolls would have been invalids most of the time if it had not been for Parthenope's little family, who often required their Aunt Florence's care'.[14] When not working in her dolls' hospital, Florence was busy delivering food, medicines and other comforts to the poor on her father's estate; her 'child-sympathy was so marked as to elicit remarks'.[15]

A favourite story was little Florence's successful nursing of the injured sheep-dog, Cap. She and the local minister, who had fortuitously trained in medicine, make the rounds of the local villagers. They find Roger, the old shepherd, unable to herd his sheep without Cap, who has been injured by unruly boys; Roger must now kill Cap, because he cannot afford to feed him. Florence and the minister hasten to Roger's cottage, where they discover that Cap only has a bad bruise: 'It was dreadfully swollen, and hurt very much to have it examined; but the dog knew it was meant kindly, and though he moaned and winced with pain, he licked the hands that were hurting him'.[16] Under the eye of the minister, Florence heats water and applies hot compresses to bring the swelling down. Only when Cap is obviously better do they leave. On the way home, they meet the shepherd.

> 'Oh, Roger,' cried Florence, 'you are not to hang poor old Cap; his
> leg is not broken at all.'
> 'No, he will serve you yet,' said the vicar.
> 'Well, I be main glad to hear it,' said the shepherd, 'and many
> thanks to you for going to see him.'[17]

The next day, Florence returned to finish her task of caring for Cap, and Roger later says, 'Do look at the dog, miss; he be so pleased to hear your

---

his vocation during his long voyage on the *Beagle*. This is never mentioned in the biographies for boys.

[13] Mabie and Stephens (1908), 'Sister Dora', 241–2.
[14] McFee (1902), 7.
[15] Thayer (1897), 33.
[16] Mabie and Stephens (1908), 269.
[17] Alldridge (1885), 14.

voice . . . I be greatly obliged to you, miss, and the vicar, for what you did. But for you I would have hanged the best dog I ever had in my life'.[18]

The above incident is almost too obviously over-determined. 'Mischievous schoolboys' – the enemies of all right-thinking girls – have committed the thoughtless crime of harming a working farm animal. The young lady-bountiful prevents a sad mistake from being committed by the peasant-shepherd, who is driven by economic necessity to destroy an animal that cannot earn its living. Both the dog and the shepherd are suitably grateful. Cap, even when in great pain, does not nip those who do him good; when he later hears Florence's voice, he wags his tail in gratitude, but without taking his eyes from the sheep he is guarding. Most accounts also add that Florence remembered to give two petticoats to the neighbour who had lent one to be torn up for compresses. Class privilege and class lines are maintained by everyone playing the appropriate role.[19]

Even more striking, however, is the audience for Florence's actions. She does not as a young girl travel alone around her father's property, but instead is accompanied by the local Anglican clergyman, that figure of Christian wisdom and status. He teaches her even as he instructs and counsels his parishioners. Moreover, his medical expertise, rather than the common sense of the experienced shepherd, leads him to 'believe that the leg is [not] really broken. It would take a big stone, and a hard blow, to break the leg of a great dog like Cap'.[20] He correctly diagnoses and prescribes for Cap; Florence follows his orders in the time-consuming preparation and application of compresses. The girl Florence thus learns to strengthen her naturally feminine propensities to nurse under the aegis of a caring male.

The device of a male observer approving the actions of the girl-heroine is repeatedly used in Victorian biographies. Often he is a sympathetic minister, teacher or relative – desexualised, even feminized, figures whom a girl was likely to know well. Suffering from her mother's disapproval of her eagerness to learn, Mary Somerville confessed her love of Latin and mathematics to an encouraging uncle. A minister, like a woman, was expected to care for and sympathise with the poor; an uncle could be concerned with a niece's future

---

[18] Ibid. Some accounts soften the economic reality of a shepherd's life by claiming that Roger intended to kill Cap because of the seriousness of the injury, rather than the economic impossibility of keeping a non-working animal.

[19] The repetition of this story for over fifty years may have been due to its similarity to the immensely popular tales of innocent waifs and dainty girls saving irreligious parents, drunken outcasts or other melodramatic figures. See Avery (1975), 112–20 and 150–5.

[20] Alldridge (1885), 12.

without worrying about discipline; even a father could bend to a child's needs in ways that a mother would not. These men, however, only guide the heroine; she herself takes the lead, following a natural, pre-existing bent – to nurse, to study or to teach. Men introduced girls to the outside world, helped them to master necessary skills, and gave them moral approval; women, such as Elizabeth Fry, served as inspiration.

An innate desire to do good and win adult approval was not sufficient for the successful nineteenth-century self-made hero or heroine. Proper training, hard work and patience were also necessary to succeed. Readers were warned that Florence Nightingale 'devoted ten years to study and training before she undertook even the comparatively small responsibility of looking after Harley Street Nursing Home and Hospital for Governesses. But when she was called to higher duties she was fit for them, and she then performed her task in a way that won the admiration of the universe, and established a precedent for all time'.[21] Given the existing state of nursing, much could be made of Nightingale's strenuous efforts to find appropriate training. Her brief stay at Kaiserswerth, where Pastor Fliedner trained working-class women in rudimentary nursing skills, and her numerous visits to continental hospitals served to highlight the need for change in England – and the achievements of Nightingale and her followers in the years following the Crimea. Yet the actual education undertaken by Nightingale, or any other nurse for that matter, was described more in terms of training in self-discipline and hard work than in actual medical procedures or technical knowledge.

Biography by its very nature emphasises the achievements of the lone individual. Individual effort is supreme. Mary Somerville struggled in isolation to learn algebra and geometry; her mother even forbade her candles when she was discovered working late at night on Euclid. Only when she was a widow with two small children, at thirty-three, could she begin a systematic study of mathematics and astronomy. Yet the modest, very feminine Somerville insisted in her autobiography that she had always met with the greatest courtesy from men in her chosen field.[22] Most heroines were portrayed, like their male counterparts, struggling against an uncomprehending world, rather than their families. Virtually no mention is made of the bitter opposition of Nightingale's family to her nursing studies.

[21] Pratt (1897), 121.
[22] These details are documented in her autobiography, edited by her daughter, Martha Somerville (1874), 16–80.

Instead, her decision to reform nursing is dramatised as an act of individual courage:

> Nursing was a base profession, not much above that of a barmaid; and Florence Nightingale was a lady born and a lady bred. She had to have the confidence that she could preserve herself from contamination while she elevated the profession. And it was a prodigious undertaking. But she was not without inspiring examples. She laid her case before Elizabeth Fry, who had renovated the prisons of Europe, and was, of course, encouraged to go ahead . . . Yet her strongest inducement was not these examples but her knowledge of the shocking need for intelligent nursing. It was, above all, her own leaning toward that humane occupation. For that was the woman's natural bent — she who had healed Cap, and read consolation to her rheumatic neighbors. She did not lack courage.[23]

The well-trained, patient Nightingale is rewarded with the high drama of fighting to save sick and wounded soldiers. But her success depended not only upon her nursing knowledge, but also her class privilege and unique feminine skills. Although none of the medical men and military officers recognised Nightingale's value at first, in a dire emergency, her cool-headed calm won the day. Through foresight, she had brought a supply of desperately needed food and clothing with her; working twenty hours a day, she was able to soothe and oversee those in danger of dying; and, relative order once established, she was ready 'to do battle with the monster Red Tape.'[24]

Another biographer concluded: 'But after all is said of Florence Nightingale's sympathy and her science, she owed her final triumph in the Crimea to a rarer talent, that of tactful organising and executive power'.[25] Girl readers were encouraged to think of themselves as possessing these specific powers, rather than having the scientific bent of their brothers. They could enter the modern world without losing their femininity if they developed self-control, rationality and organisational skills.

The description of Nightingale's reformed hospital is reminiscent of the

---

[23] Adams and Foster (1913), 128–9.

[24] Richards (1920), 94. It is interesting to note the use of the slang 'Red Tape.' Nightingale referred to her chief enemy, Dr (later Sir) John Hall, the Inspector-General of Hospitals in the Crimea, as 'the prince of Red Tape and inhuman routine'. See Goldie (1987), 200.

[25] Adams and Foster (1913), 138.

well-run country house from which Nightingale had come. The rigid discipline and class hierarchy of the domestic sphere is replicated in the reformed hospital:

> Miss Nightingale's headquarters . . . was a large, airy room . . . in the middle was a large table, covered with stores of every kind, constantly in demand, constantly replaced. . . . Bales of shirts, piles of socks, slippers, dressing gowns, sheets, flannels – everything you can think of that is useful and comfortable in time of sickness. About these piles the white-capped nurses came and went, like bees about a hive; all was quietly busy, cheerful, methodical. In a small room opening off the large one the Lady-in-Chief held her councils with nurses, doctors, generals or orderlies; giving to all the same courteous attention, the same clear, calm, helpful advice or directions. Here, too, for hours at a time, she sat at her desk, writing; letters to Sidney Herbert and his wife; letters to Lord Raglan, the commander-in-chief, who, though at first averse to her coming, became one of her firmest friends and admirers; letters to sorrowing wives and mothers and sisters in England.[26]

Richards placed Nightingale's leadership and adminstrative skills within a world of socially inferior men; by a combination of will power, authoritative demeanour and womanly calm, Nightingale succeeded in bringing about administrative efficiency and order. The medical men learn to admire and assist her in her battle against military Red Tape. She could dominate the chaos of the hospital because of her natural superiority, backed by her long training in self-discipline and determined acquisition of nursing skills.

One of the less obvious reasons why Nightingale was so attractive a heroine to girls was her patriotism. In a society that valued imperial warfare as a testing ground for young men, her war work showed how women too could help the nation. The successful care of the Crimean soldiers quickly came to epitomise patriotism with a feminine face. Under both adversity and adulation Nightingale never lost her essential womanliness. Even as she became England's best loved war heroine, she accepted the nation's praise in an especially womanly manner: 'the English people' had planned 'a public welcome of their heroine, but with the modesty and calm judgement that

[26] Richards (1920), 100–1.

always characterised her, she slipped quietly into England by the carriage of a French steamer and so to her country home'.[27]

A single description of the impact of the nurses — and not just of Nightingale — demonstrates the power that women were given, as well as what they might seize:

> The most hardened soldier could not be indifferent to such self-abnegation. The bare thought of such self-sacrifice by a woman to mitigate his trials would inspire him to appear at his best. The presence of women, as nurses, amidst the horrors of war, could not be otherwise than elevating and refining, begetting patience, contentment, and courage. For heroines to come to the aid of distressed soldiers must make heroes of them. They could not be weak, unmanly, and dispirited, with such examples of self-denial and Christian sympathy moving up and down the wards daily.[28]

Women are placed in their traditional role of inspirer-of-men, but something more is added. They alone can change the hospital. Male space — the military hospital — has been successfully invaded and morally and physically revolutionised. Female morality, when backed by modern administrative methods, is more powerful than male aggression. The women are first heroines, before the men can be reformed — they are the leaders of men who are 'weak, unmanly, and dispirited'. Heroines, not wars, create heroes.

With the help of her nurses, Nightingale is the active agent who brings order out of military chaos, but the means by which she does so must be suppressed, lest it prove to be unfeminine. Like the tales of converting the natives, where the process of bringing hundreds to Christianity is silenced in favour of the end result, all heroines bring domestic order, religious belief or physical safety out of public chaos by sheer force of will. In order to avoid such complications as ambition, cunning, manipulation and possibly even dishonesty, the narrative slips over the tiresome details of Nightingale's actual daily behaviour to the end result of her work. A few examples of washing patients or insisting on medication for the dying speak for the whole. Or, a specific event will mythologise the heroine. Nightingale's silent, nightly rounds through the miles of wards, when the waiting soldiers were said to kiss her passing shadow, transforms her into the 'Lady with the

---

[27] Mabie and Stephens (1908), 277.
[28] Thayer (1897), 52.

Lamp'. Metonymy serves to cover the complexities of achieving social change.

But metonymy and narrative gaps also leave room for the reader to insert herself. Too many uncomfortable details left little to the imagination. If formulaic biographies were to work successfully, they had to permit the reader imaginative escape into the life of the heroine. The reader, perhaps hopeful of a different, wider life, could interweave her inchoate dreams with the narrative. The simplified actions of the heroine made identification easy, regardless of how remote they might be from the reader's life. A shadowy Nightingale could be invested with desirable characteristics more easily than one encumbered with complex motivations. The creation of vivid symbolic behaviour actually counteracted the tendency of the narrative to reify women. Even though the overt message was often womanly obedience, symbolic moments, such as Nightingale's nightly vigil, permitted an imaginative identification with independent action. Fiction – created by the reader and not just the author – was an essential part of the biography.

A personal heroine for modern girls

If we look to later biographies of Florence Nightingale, a number of interesting alterations occur, all of them, however, confirming the importance of innate feminine characteristics for public heroism. Each reflects the changing social attitudes toward women; none focuses upon the medical or scientific aspects of nursing or nurses' training. Whatever modernisations were added to the biography, it remained a tale of individual achievement and the overcoming of social opprobrium in order to bring about the reform of a woman's occupation. Although personal details increase, innate feminine characteristics dominate.

A major difficulty faced by biographers after Nightingale's death was the extraordinary increase in information about her. By the interwar years, Lytton Strachey's *Eminent Victorians* (1918) and Sir Edward Cook's definitive biography (1914) had diminished Nightingale's reputation. Her name appeared less frequently on the lists of famous women.[29] But in the 1950s, Nightingale reappears, suitably psychologised. During the reign of the 'feminine mystique', a girl's ambition had to be carefully controlled. Florence and

---

[29] I have found only three biographies for girls: Tabor (1925), Willis (1931) and O'Malley (1933). O'Malley is based on her biography for adult readers (1931).

her sister Parthenope were no longer described as both interested in helping the poor; rather, Parthenope was 'entirely different' from her in her love of frivolity and easy living. Nightingale sounds like a typical idealistic adolescent in Yvonne ffrench's description:

> She had a precise, practical mind, and passionately intense feelings. She was morbid, she was self-willed, and a real problem to her parents who could not understand why she should not be as happy and contented as her sister.
>
> She became thoroughly out of sympathy with her family, her relations and their attitude to everything. Even before she was a fully grown woman she was conscious of being set apart. She felt that she had a mission to do good in the world; she felt in some way called to a dedicated life, but what that was to be she did not know. It worried her.[30]

To appeal to the modern girl reader, Nightingale's family trouble replaces Roger and Cap. The result is an attractive modernising of an old story, but also the loss of the palpable lesson of public duty taught by nursing Cap.

The mid twentieth-century focus upon the temptations of material wealth and family quarrels subtly undermined the arguments for individual initiative and responsibility taught by the earlier versions. Moreover, the unspoken space between initial efforts and final success that had characterised Victorian biographies now became an insurmountable gap. A girl heroine who was busy fighting with her parents and giving up balls for sick neighbours might have been more accessible than the austere heroines of the past, but she was also less capable of independent action. Indeed, ffrench credited Richard Monckton Milnes, the man who loved Nightingale in her twenties, with influencing her to take an interest in the poor. Romance, rather than early childhood inclination, became the impetus for doing public good. The incorporation of sex, however muted, into the formula biography shifted the plot from a girl's negotiation between self and the public world into a choice between private romance and public service.

Biographies could keep their particular psychological hold over a youthful readership so long as they continued to create independent heroines who were guided but not led by asexual men. When they began to add romance, and especially the giving up of romance, they failed. Janice Radway has

---

[30] ffrench (1954), 8.

pointed out the failure of popular romances that create emotional triangles; her romance readers overwhelmingly preferred plots that show the gradual removal of the barriers separating a woman and a man, rather than those that turn upon the winning of the beloved in competition with another person.[31] Her insight, I believe, applies equally to girls' biographies, which had traditionally based upon the opposition of two forces and not three. When the confrontation between the heroine and an uncomprehending society was changed to a struggle within to choose between a man and reforming society, the biography was no longer the story of a successful public heroine, but rather of a failed romantic heroine.

The helpful male guide cannot be sexualised without the dramatic loss of psychological independence for the girl. In her 1954 biography Yvonne ffrench describes vividly the struggle Nightingale underwent before becoming a nurse:

> Gradually, very gradually a slow change in herself was being
> effected. A hardening process was beginning, and dedication to a
> cause replaced devotion to friendship. A deliberate renunciation of
> her friends gradually followed, and in her private notes she made
> the final act of self-denial: 'Oh, God, no more love. No more mar-
> riage. Oh God.'[32]

ffrench is truer to Nightingale's actual experience, but what young woman would want to choose public service after reading about Nightingale's painful sacrifice of the sympathetic and helpful Richard Monckton Milnes? Who wants to struggle vicariously with Nightingale's dilemma? When personal love so conflicts with heroism, the preferred choice is obvious.

More recent rewritings of Nightingale's story have not been especially successful in escaping the well-established formula of girls' biographies. Robin McKown, writing in 1966 at the height of the Vietnam war, and on the eve of the second wave of feminism, includes an epigraph that echoes numerous similar comments made during the years following the Crimea:

> Nothing can replace a woman at the bedside of a wounded man,
> not only at base hospitals, but more especially at the front. The
> year before, when a wounded man arrived at my field hospital,
> scared and in pain, he immediately found two girls dressed in white

---

[31] Radway (1984), 171–2.
[32] ffrench (1954), 17.

leaning over his bed. That was enough: his doubts and anxieties were over there and then. He felt he was safe the moment he saw a girl's quiet smile. After all, one has to smile back at an attractive face — it's only polite — and one has to brazen it out a bit too and pretend there's nothing wrong even when one is ready to howl with pain. One has to show gratitude for so much gentleness and patience, for the mysterious feminine quality which appears even in the most professional of actions. It acts like a charm.[33]

McKown, like ffrench, mentions Monckton Milnes, but as befits a book written during a war, most of her nurse heroines had worked in wartime conditions. She focused on Nightingale's administrative skills in bringing order to the filthy and chaotic hospitals, which are described in detail. Aside from the updated vocabulary, the image of Nightingale is reminiscent of biographies written nearly a hundred years earlier: 'Using tact, diplomacy and feminine wiles, she gradually imposed her will on the fearful or callous hospital officials. Her manner was both serene and firm . . . It was amazing what this slender, delicate woman accomplished'.[34]

McKown and other recent biographers have connected Nightingale with the first American woman doctor, Elizabeth Blackwell, rather than the Quaker philanthropist, Elizabeth Fry. They see more affinity between a pioneering professional whom Nightingale's mother deeply distrusted than an untrained, religious volunteer. Beatrice Seigel in a 1991 biography even includes a paragraph on Barbara Smith Bodichon, the most important of the early British feminists and a first cousin of Nightingale's.[35] Although Nightingale did not know her radical cousin, Seigel is as anxious to give her a feminist context as an earlier generation was to give her orthodox femininity.

Nightingale may be unrecuperable as a girl heroine in the late twentieth century. We now know almost too much about her — her unpublished papers constitute one of the largest single collections in the British Library; over 13 000 combative letters remind us of her manipulative, aggressive ways. Ever since Lytton Strachey's demystification of the Victorians, scholars have been trying to dislodge the popular image of 'the Lady with the Lamp'. F. B. Smith, in a sustained attack on 'her talent for manipulation' considers her to be a hopeless adminstrator who alienated all those who worked under

[33] Major Paul Grauwin, *Doctor at Dienbienphu*, quoted in McKown (1966), 8.
[34] McKown (1966), 65.
[35] Siegel (1991), 46. Nightingale was never permitted to meet her first cousin because of her uncle's unorthodox common-law marriage.

or with her. He describes her brief sojourn at Harley Street Institution for the Care of Sick Gentlewomen as a virtual failure, but 'by her insouciant explanations of her failures with staff, patients, tradesmen and accounts and her generalising of the blame . . . biographies have been beguiled ever since. Her superbly assured epigrammatic Byronic prose by a process of stylistic legerdemain turns small gains with linen and jellies into mighty personal feats and big set-backs into everybody's moral shortcomings. By style and instinct she was a consummate politician'.[36] Like all modern biographers, Smith has no truck with her so-called nurses' training or her scientific knowledge, though he does acknowledge her intelligence.

Nursing historians have found Nightingale an especially difficult incubus, associated as she is with selfless service. In an occupation determined to upgrade its public image, professional status and salaries, her legacy is at best mixed. Feminist historians have also found Nightingale a generally unattractive figure. While they have been more willing than F.B. Smith to grant her considerable administrative expertise, her flagrant use of class privilege, consistent refusal to recognise the capacities of other women, and recalcitrant unwillingness to accept antisepsis, the germ theory and virtually all the other medical discoveries made in her lifetime, make her a tough heroine to admire.[37] Under the circumstances it is hardly surprising that the Feminist Press chose to publish a biography of Elizabeth Blackwell rather than the far more famous Nightingale in its series of revisionist stories for girls.

What elements, then, made Nightingale the single most popular heroine in girls' biographies of the late nineteenth and early twentieth centuries? While the overt message of so many biographies may have been hard work and obedience, the extraordinary adventure in the Crimea did not encourage quiescence. Even as biographies purported to teach modesty and home duties, they awakened an imaginative affinity with active, patriotic heroines. Moreover, since these Victorian heroines were not invested with personal or sexual conflicts, they widened the definition of the possible. Asexual heroines could be admired for their intrinsic qualities, however stereotyped these may have been. Nightingale had a glamour and privilege that could not necessarily be emulated, but her lofty self-control and influence over others

---

[36] Smith (1982), 17. See also the more judicious contextual study by Rosenberg (1979), 116–36.

[37] Baly (1986) documents Nightingale's complicated relationship with both the Nightingale Fund, raised by the people of England in thanks for her Crimean War work, and the Nightingale School of Nursing, founded in 1861 with these funds. Summers (1988) documents the positive efforts of other nursing groups in the Crimea and Nightingale's bitter opposition to them. See also the relevant essays in Davies (1980), Holton (1984), Poovey (1988) and Whittaker and Oleson (1978).

was a potent example of feminine power. The very characteristics that we now suspect – her rigid authoritarianism, appeals to faith, and shameless use of others for a higher purpose – were precisely what most attracted her contemporaries. Unlike so many pioneering women scientists and doctors, those who imitated Nightingale were not deluded by notions of intellectual equality and scientific objectivity.[38] They gloried in women's special powers, for who had been more effective in using them for change than Nightingale?

Acknowledgements

Portions of this chapter have appeared previously in a different form in 'What makes a heroine? Nineteenth-century girls' biographies', *Genre*, **20**, (Summer 1987), 171–80. Thanks to the editors for permission to reprint. Special thanks once again to Barbara Sicherman, who suggested sources and critiqued different versions of this chapter. Her knowledge and analysis of the reading material of girls and women has been invaluable.

Bibliography

Abir-Am, P.G. and Outram, D. (eds) (1987) *Uneasy Careers and Intimate Lives: Women in Science, 1789–1979*. New Brunswick: Rutgers University Press.

Adams, E.C. & Foster, D.F. (1913) *Heroines of Modern Progress*. New York: Sturgis and Walton.

Alldridge, L. (1885) *Florence Nightingale, Frances Ridley Havergal, Catherine Marsh and Mrs Renyard*. London: Cassell.

Avery, G. (1975) *Childhood's Pattern: A Study of the Heroes and Heroines of Children's Fiction* London: Hodder and Stoughton.

Baly, M.E. (1986) *Florence Nightingale and the Nursing Legacy*. London: Croom Helm.

Bolton, S.K. (1926) *Famous Men of Science*, revised edition. New York: Thomas Y. Crowell.

Chappell, J. (1898) *Four Noble Women and their Work*. London: S.W. Partridge.

Davies, C. (ed.) (1980) *Rewriting Nursing History*. London: Croom Helm.

ffrench, Y. (1954) *Florence Nightingale, 1820–1910*. London: Hamish Hamilton.

Glazer, P.M. and Slater, M. (1987) *Unequal Colleagues: the Entrance of Women into the Professions, 1890–1940*. New Brunsick: Rutgers University Press.

Goldie, S.M.(1987) *'I have done my duty': Florence Nightingale and the Crimean War*. Iowa City: University of Iowa Press.

Haydon, A.L. (1909) *Florence Nightingale, O. M.: a Heroine of Mercy*. London: Andrew Melrose.

Holton, S. (1984) Feminine authority and social order: Florence Nightingale's conception of nursing and health care. *Social Analysis*, **15**, 59–72.

---

[38] See Glazer and Slater (1987), 80–93, for a discussion of the pitfalls of assuming intellectual equality in a scientific world without the support of a wider feminist community.

[Johnson, J.] (1863) *Clever Girls of Our Times: and How they Became Famous Woman.* London: Arton and Hodge.

[Lockyer, N.] (1870) Scientific administration. *Nature*, 2, 449.

Mabie, H.W. & Stephens, K. (eds.) (1908) *Heroine Every child Should Know.* New York: Grosset and Dunlap.

McFee, I.N. (1902) *The Story of Florence Nightingale.* Dansville, New York: F.A. Owen.

McKown, R. (1966) *Heroic Nurses.* New York: G.P. Putnam's Sons.

O'Malley, I.B. (1931) *Florence Nightingale 1820–1910: a Sketch of Her Life Down to the End of the Crimean War.* London: Thornton Butterworth.

O'Malley, I.B. (1931) *Great Englishwoman: Florence Nightingale.* London: G. Bell and Sons

Poovey, M. (1988) *Uneven Developments: the Ideological Work of Gender in Mid-Victorian England.* Chicago: University of Chicago Press.

Pratt, E.A. (1897) *Pioneer Women in Victoria's Reign.* London: George Newnes.

Radway, J. (1984) *Reading the Romance: Women. Patriarchy, and Popular Romance.* Chapel Hill: University of North Carolina Press.

Richards, L.E. (1920) *Florence Nightingale The Angel of the Crimea: A Story for Young People.* New York: D. Appleton.

Rosenberg, C.E. (1979) Florence Nightingale on contagion: the hospital as moral universe. In C.E. Rosenburg (ed.) *Healing and History.* New York: Dawson.

Rossiter, M.W. (1982). *Women Scientists in America, Struggles and Strategies to 1940.* Baltimore: Johns Hopkins University Press.

Salmon, E. (1886) What girls read. *Nineteenth Century*, 20, 515–29

Schiebinger, L. (1989) *The Mind Has No Sex? Women in the Origins of Modern Science.* Cambridge, MA: Harvard University Press

Siegel, B. (1991) *Faithful Friend: The Story of Florence Nightingale.* New York: Scholastic Press.

Smith, F.B. (1982) *Florence Nightingale: Reputation and Power.* London: Croom Helm.

Somerville, M. (ed.) (1874) *Personal Recollections, from Early Life to Old Age of Mary Somerville.* London: John Murray.

Summers, A. (1988) *Angels and Citizens: British Women as Military Nurses, 1854–1914.* London: Routledge & Kegan Paul.

Tabor, M.E. (1925) *Pioneer Women: Florence Nightingale.* London: Sheldon Press.

Thayer, W.M. (1897) *Women Who Win or Making Things Happen.* London: T. Nelson and Sons.

Vicinus, M. and Nergaard, B. (1989) *Ever Yours, Florence Nightingale: Selected Letters.* London: Virago.

Vicinus, M. (1985) *Independent Women: Work and Community for Single Women, 1850–1920.* Chicago: University of Chicago Press.

Whittaker, E. and Oleson, V. (1978) The face of Florence Nightingale: functions of the heroine legend in an occupational sub-culture. In R. Dingwall and J. McIntosh (eds) *Readings in the Sociology of Nursing.* Edinburgh: Churchill Livingstone.

Willis, I.C. (1931) *Florence Nightingale: a Biography for Older Girls.* New York: Coward–McCann.

8

Taking histories, medical lives: Thomas Beddoes and biography

ROY PORTER

Medical biography

The prospects and problems of scientific biography have been extensively discussed in recent years; creativity has been probed and analyses offered of the historical tensions between the private pursuit of science and its public face.[1] If the career of a scientist implants expectations that the biography itself will enshrine 'scientific' virtues – factual accuracy and objectivity – the very opposite might be suggested by study of the lives of psychoanalysts. To plumb the murky depths, would not an analyst's biographer need to investigate the unconscious, and thereby advance inferences subjective and unfalsifiable, leading to unfathomable transferences between biographer and biographee? Doubtless tongue in cheek, Freud himself foisted a totally deed-oriented deadpan autobiography upon the world, but later non-Freudian accounts of the father of psychoanalysis, notably Ronald Clark's *Freud: The Man and the Cause* (1980), disappoint because they do little more than skim the surface.[2] If physicians may be situated, vocationally and perhaps temperamentally, somewhere between scientists and psychoanalysts, does that mean that there is, or ought to be, a unique mode of *medical* biography, to suit the distinctive qualities of their subject? Is there a distinctive genre of medical biography? If not, *should* there be? These are questions I wish to pose in this chapter. Lest I raise expectations that will be disappointed, let me say from the outset that I cannot pretend to offer any very general or

---

[1] For a sample of such musings see Holmes (1981), 60–70; Hankins (1979); Sheets-Pyenson (1990); and, perhaps more surprisingly, Young (1988).

[2] See Clark (1980), and, for evaluation, Young-Bruehl (1994). More broadly on the strengths and weaknesses of psychohistory see Gay (1985); Loewenberg (1983); DeMause (1975); Stannard (1980). The interplay of films about Freud with Freudian films has been explored in Gabbard and Gabbard (1987) and Shortland (1987).

theoretical resolutions. Such a notion would be unrealistic, I believe, in view of the enormous historical diversity of the practice of medicine. I do want to suggest, however, more through the concrete instance discussed in the body of the paper than by way of abstract propositions, that the medical clinician is himself, in large measure, necessarily a biographer – of his patients, examining them as complex individuals within a social milieu. Perhaps that makes the biographer a kind of vicarious doctor. One of the ways, therefore, that medical biography may be specially illuminating is precisely because it is relentlessly and inescapably reflexive. I do not mean by this that the biographer, in some literal-minded way, necessarily finds out about *himself*. But, if my own limited experience of having written one small medical biography may be anything to judge by, the attempt to write the life of a doctor does, I believe, inevitably provoke deliberations about what it is to 'take a history' or know ('diagnose') a character on the basis of limited evidence (or symptoms).

Medicine was conventionally a profession sustained by ancestor worship. Young doctors were trained to venerate noble forefathers, Hippocrates and Galen, Harvey and Sydenham, and medical history was served up doxographically, a succession of portraits of heroic discoverers and sagacious clinicians. From Samuel Johnson to Sir William Osler, medical biography was written in a pious, uplifting tone. The lives of the doctors were valuable for what they could teach.[3]

Exemplifying what J.H. Plumb has dubbed 'the death of the past',[4] modern medicine is perhaps losing interest in its ancestor cult, and professional medical historians have certainly been primed to avoid hero-worship. The result is that, though biographies of doctors proliferate, it is remarkable how many eminent physicians still lack a scholarly modern 'life', and equally noticeable how few truly significant medical biographies have been written by leading academics. To take my own special field of interest, Britain during the 'long eighteenth century', it is astonishing that there has been no life (of any kind or quality) of William Cullen since the 1830s, no modern study of John Brown, no full-scale account of John Fothergill; the last biography of John Coakley Lettsom is sixty years old, the first life of William Heberden has only just appeared.[5]

---

[3] For some discussion, see Brieger (1980, 1993); Bynum (1980); Pelling (1983); Webster (1983). For Johnson, see Dusseau (1979); McHenry (1959). For Osler, see White (1939); Kingsbury (1987). For a source see Morton and Moore (1989).

[4] Plumb (1969).

[5] See Abraham (1933); Heberden (1990). For Fothergill see Corner and Booth (1971); Fox (1919), and Booth (1987). The most recent biography of Brown was Brown (1903).

Medical biography doubtless poses tantalising problems. In many cases, hardly any private papers survive.[6] Physicians and surgeons have commonly been authors; but healing is preeminently an *activity*, involving routine practices and standardised therapeutic procedures. Details of these may have been jotted down in doctors' pocket-books, but few of these have survived the wreck of time. The same applies to case-notes; and in any event, being formulaic, case-records typically say less about the individual physician than about medical protocols.[7] Above all, bedside medicine is so intimate, so much bound up with manner, personality and a ring of confidence – ephemeral, charismatic features that defy being written down – that it is vastly difficult for the biographer to document the day-to-day, experiential reality of the healing art, perhaps as forlorn a task as that of the music critic, compelled to freeze performance into prose. That is why the genre of the medical anecdote is so revealing and yet so frustrating: stylised and stereotypical, it is often wanting in the veridical, but the *ben trovato* may capture supremely well certain folkloric and dramatic elements of the medical art.[8] The quandaries of medical biography permit no simple resolutions, and the business of grasping *la vie médicale* requires broader discussion than can be tendered here.[9] I do, however, wish to float a particular proposal, which I shall follow up with a case-study.

One possible *point d'appuie* for probing the lives of certain medical men, I suggest, lies in exploiting the affinities between medicine and writing. Extensively explored of late, these parallels need not be rehearsed here. 'Taking a history', after all, is central to the crafts of physicians, novelists and biographers alike. Doctors and story-writers both explore individuals, diagnosing complaints and abnormalities, and tracing them back to some source. All interpret disjointed evidence, trying to translate fragments into a

---

6 An instance is the doctor-*philosophe*, La Mettrie. Wellman (1992) is severely hampered by this lacuna. For a good instance of what may be accomplished in the absence of substantial MS materials, see Wilson (1992). In other cases, papers survive till now little utilised. Helen Brock is currently producing an edition of William Hunter's letters, which run to some 700. See Brock (1983) and Bynum and Porter (1985). W. F. Bynum, Roy Porter and Andrea Rusnock are editing the recently discovered correspondence of James Jurin, physician, promoter of smallpox inoculation and secretary of the Royal Society. Though these letters run to over 700, it would be idle to pretend that they afford a rounded or intimate picture of the man.

7 William Heberden's case-registers are quite personally revealing: see Heberden (1990). The early eighteenth century German physician, Johannes Storch, kept superb records; their interest, however, is representative and typical rather than personal: see Duden (1987). For the genre, see Geyer-Kordesch (1994).

8 An ethno-methodologic study of medical anecdotes is a desideratum. For compilations of anecdotes see, for instance, *Anecdotes Medical, Chemical and Chirurgical; Collected, Arranged and Transmuted by An Adept* (1816); Heathcote (1786); Andrews (1896); Jeaffreson (n.d.); Timbs (1876).

9 For an early tradition of medical and scientific biography, the *éloge*, see Outram (1978); Weisz (1988); Paul (1980).

complete picture. In each, writing is involved — even if, with the clinician, this may routinely involve little more than scribbling 'magic' prescriptions. Writers and doctors both create authority, and each is involved in activities designedly therapeutic — the physician seeks to heal his patient, the author to produce catharsis, to instruct, divert and restore.[10]

It is no accident that so many medical men have doubled as authors — in eighteenth-century Britain Locke, Mandeville, Akenside, Smollett, Goldsmith and Erasmus Darwin, to name but a few. And it is no surprise that many of these, notably Mandeville, have harped upon the parallels between writing, bodily processes, sickness, and the healing arts.[11] Satirically or philosophically, moralistically or psychologically, certain doctors have mused on what was involved, in Sterne's disturbing phrase, in 'taking a man's character', and by consequence have been self-reflexive about their role as 'prescribers' in the human tragi-comedy. Sensitised to both the 'sick role' and the 'healer's role', the writer/doctor thereby presents special challenges but, for that reason, golden opportunities for the medical biographer. In the pages to follow, I wish to trace and elaborate some of these themes through the career of one late eighteenth-century physician, Thomas Beddoes (1760–1808). He is an apt exemplar. He was an energetic clinician but also a prolific author in a number of genres on a spectrum of subjects, from medicine to logic and geology. In his occasional and popular writing, he conjured up before the reader's eye scenes featuring doctors and patients, sometimes didactically, sometimes ironically. Not least, he diagnosed the health risks, psychological and physical, of the innovative high-pressure, commercial, Enlightenment society, whose information and communications revolutions (the birth of the novel!) were driving the polite mind to the point of perilous overload.

Beddoes offers a fascinating example of a physician aware of new psychological epidemics. Yet his biographer also faces stubborn problems. Soon

---

[10] These concerns are central to American 'medical humanities'. See, *inter alia*, Rousseau (1991*a*); Trautmann (1982); Trautmann and Pollard (1975); Cousins (1982); Brody (1987); Morris (1991). For discussions of eighteenth-century doctor/writers, see MacNeil (1987); King-Hele (1977); Rousseau (1991*b*); Porter (1989*a*); Benjamin (1990); Porter (1992*a*). For writing and sickness see Cousins (1982); Leader (1991); Porter (1989*b*). See also the 'Introduction' to Roberts and Porter (1993), 1–22, where these matters are discussed at length; this also contains an extensive bibliography on approaches to the understanding of the image of the physician and the interfaces between writing and healing. For similar matters, see also Peterfreund (1990); Levine (1987).

[11] See especially McKee (1991). The playwright offering his work as a healing entertainment, the philosophical moralist presenting state physic, formed common tropes: see Silvette (1967), and Seymour (1992); and see the exploration of similar metaphors in the 'Introduction' to Porter and Porter, (1993).

after his death, his widow, Anna commissioned a highly doctored biography of her husband and then, it seems, had most of his papers destroyed (she clearly feared she had much to hide: she was in love with her husband's best friend). Beddoes's life thus comes to us at several removes.[12]

Through exploring this doctor's attitude to 'writing the self', I wish to raise the following questions, which, hopefully will be of more representative interest. How did Beddoes's dual identity as doctor and writer impact upon his perceptions of the sources of sickness in others? How did his sense of professional self shape his perceptions of his inner self?[13] How did he assess the relations between writing, reading, sickness and the self?[14] How should historians assess his 'project'? First, however, it will be illuminating to give a thumbnail sketch of his life and works.

Or will it? Because the notion of such a vignette presupposes some 'objective' truth of life, a factual foundation upon which an edifice of interpretation may be built. Yet such a notion is at best tendentious and at worst fallacious. Beddoes lives in the eye of the beholder. His first and 'official' biographer, J.E. Stock, chose, for reasons totally intelligible in respect to the family and the times, to downplay Beddoes's political involvements and his radicalism. His most thorough biographer, Dorothy Stansfield, gives generous coverage to his scientific researches, but says far less about his medical activities. I happen to see, and to have written about, a rather different Beddoes: a man obsessed by the politics of medicine, the maladies of the body politic, and the vision of the statesman as state physician. These roles also, in my view, shed great light on the inner man. My sketch is a silhouette of Beddoes that depicts these outlines.

Doctor of revolution

Thomas Beddoes was born into a prosperous Shropshire tradesman's family in 1760, somewhat junior to the luminaries of the Lunar Society, but senior to the great Romantics — Wordsworth, Coleridge, Southey, Shelley, Byron and Hazlitt.[15] A tanner's son, Beddoes typified the West Midlands character:

---

[12] Stansfield (1984); Porter (1991*a*).

[13] On the fabrication of the self see Cockshut (1984); Spacks (1976); Flynn (1991); Greenblatt (1980).

[14] Practically all the following biographical information may be found in Stansfield (1984); much was first made public by Beddoes' first biographer, Stock (1811).

[15] Some introduction to the late eighteenth-century intelligentsia may be found in Schofield (1963); Butler (1981); Halévy (1924); Cone (1968); Williams (1958).

thrusting, businesslike and ambitious.[16] Aiming to rise by his own energies and industry, and hostile to tradition and privilege, at heart he was a tenacious liberal individualist, a typical bourgeois radical, dreaming of a society with room at the top.[17]

Beddoes proceeded in 1776 — that *annus mirabilis!* — to his local university, Oxford. Unlike Gibbon, he capitalised upon the true opportunities afforded by that slumbering institution, securing the comprehensive grounding in languages and literature that later enabled him to shine as a sophisticated polymath. Unusually for an Oxford man, Beddoes also cultivated scientific interests. Opting for medicine, on obtaining his B.A. in 1781 he abandoned Oxford, whose medical school was comatose,[18] moving to London, Britain's premier centre for practical medical education, and becoming a pupil of John Sheldon, successor to the celebrated William Hunter at the unrivalled Great Windmill Street Anatomy School. Three years later he proceeded to Edinburgh to complete his medical education.[19] His best experience in Scotland lay in forging a lasting friendship with Joseph Black, who, combining medical interests with international eminence as an innovator in chemistry, naturally became Beddoes's hero.

In 1786 he returned to Oxford, taking his M.D., and throwing himself into research. The Oxford chemistry laboratory afforded an ideal milieu for experimentation; no one hindered the solitary labours of the dynamic young man, while there were enough supportive colleagues to preclude isolation. Beddoes soon launched into lecturing, reviving university interest in the science. His effervescent, Romantic enjoyment of Nature also found outlet in geological expeditions around his native Welsh borderlands.[20]

Ever energetic, Beddoes widened his horizons and started to publish, bringing out, while still in his twenties, translations of eminent Continental naturalists and experimentalists: Spallanzani, Scheele and Bergman.[21] His scientific circle grew; warm friendships sprouted with Lunar Society members like Erasmus Darwin[22] and, slightly later, with James Watt. And in a bold move, in 1787 Beddoes took a summer jaunt to France, visiting Guyton de Morveau, and thereby signalling eager espousal of the new French chemistry.

[16] John Money (1977); Musson and Robinson (1969).
[17] For such political and social values, see Goodwin (1979); Thompson (1963).
[18] Webster (1986).
[19] Lawrence (1984); Rosner (1991).
[20] Porter (1977), 141, 144, 175.
[21] Beddoes (1789).
[22] Beddoes's relations with Erasmus Darwin merit more study. Desmond King-Hele has entitled his latest biography of Darwin, *Doctor of Revolution: The Life and Genius of Erasmus Darwin* (1977). The epithet seems more applicable to Beddoes.

The late 1780s found Beddoes, then breasting thirty, at the peak of his potential: he was already an internationalist and an experimenter; he had made contact, friends even, with the most productive men in the field. He had availed himself of the extraordinary openness of late Enlightenment culture. The young scientist naturally welcomed the French Revolution. Nothing could better have expressed his aspirations than the storming of the Bastille. Down with tyranny, oppression, priestcraft! A new age was at hand of liberty, the rights of man, and government by the people. At last the French were gaining what had long been the birthright of the English; would not their courage in turn promote progress at home – political, intellectual, and personal? From 1789, Beddoes speedily grew politicised: his letters to his friend, Davies Giddy, show him sporting a *tricolor*, singing revolutionary songs, and cheering universal liberty.[23] The libertarian sunrise soon turned tempestuous. With the anti-revolutionary groundswell growing in Beddoes's *alma mater*, his radicalism attracted enemies.[24] The Birmingham Riots of July 1791, in which loyalist Church and King mobs razed the library and laboratory of Joseph Priestley while the authorities egged them on, must have shocked him, for Beddoes was a younger Priestley. Though he did not know it, the Home Office was having him observed.

Beddoes took stock. Staying with his friend, the iron master, Reynolds, he penned an epic poem, *Alexander's Expedition down the Hydaspes & the Indus to the Indian Ocean* (1792), a meditation on destructive and cleansing violence.[25] At the same time, and marking his concern for popular welfare, there appeared *The History of Isaac Jenkins, and of the Sickness of Sarah his Wife, and their Three Children* (1792), an improving moral fable warning of the evils of alcohol.[26] And he also spelt out his philosophy of popular education in *Extract of a Letter on Early Instruction, Particularly that of the Poor* (1792), passionately advocating direct methods of instruction.[27] These literary writings were, in one sense, a sidestep from politics. Yet Beddoes had become a political through and through, and even these served as vehicles for political rage. *Alexander's Expedition* decried imperialism and militarism,

---

[23] This correspondence is unpublished, and is located in the Cornwall Record Office in Truro. For Giddy see Todd (1967).

[24] Always headstrong and irascible, as early as 1787, Beddoes ventured to print a fierce attack upon the mismanagement of the Bodleian Library: *A Memorial Concerning the State of the Bodleian Library, and the Conduct of the Principal Librarian. Addressed to the Curators of that Library, by the Chemical Reader* (1787). Before the 1790s, Beddoes could indulge his wrath without penalty to himself. Things changed.

[25] Beddoes (1792*a*).

[26] Beddoes (1792*b*).

[27] Beddoes (1792*c*).

while his educational writings indicted traditional schooling as the instrument of ecclesiastical obscurantism. In the best radical *philosophe* tradition, he dubbed religion a manmade monstrosity, while *Isaac Jenkins'* jeering at squires and parsons reminds one that Beddoes was Cobbett's contemporary.

Before he could be expelled, Beddoes quit Oxford in 1793, migrating to Bristol, and setting up in medical practice in the fashionable suburb of Clifton, rather touchingly in Hope Square, and, shortly afterwards, marrying Anna, the daughter of his Lunar Society friend, Richard Lovell Edgeworth.[28] Commercial Bristol was an attractive prospect, since it had opulent patients while possessing a fine artistic and literary culture.[29] Beddoes found himself in a hotbed of intellectual radicalism. For some time he had shared the radical opinions of the liberal intelligentsia: Price, Priestley, Horne Tooke, Thelwall. Now he came into direct contact with local activists. The years from 1794 to 1797 saw him ardently involved in political campaigning, journalism and pamphleteering with Thomas Poole, the publisher Joseph Cottle and, above all, with Robert Southey and Samuel Taylor Coleridge.[30]

These were heady, but also terrible times. From late 1792, Beddoes grew uneasy about the Revolution itself. In letters to his old collegian, Davies Giddy, he deplored the Terror, denouncing the 'infernal Club of Jacobins' as bloodthirsty fanatics who were wrecking the Revolution. Beddoes's radicalism perforce turned defensive — defending supporters of the rights of man even though he deplored the politics of the guillotine. English enemies of the Revolution were unquestionably his enemies, and so he denounced government policy at home, with its suspension of *Habeas Corpus* in 1793 and the Two Acts (the 'Gagging Acts') of 1795, restricting freedom of speech and assembly. In a flurry of anti-Pitt pamphlets, some positively sinister in tone — *Alternatives Compared: or, What Shall the Rich Do to be Safe?* (1797) — Beddoes won his spurs as a political writer.[31]

The early Bristol years saw Beddoes at the peak of his politicking. For a few years, Beddoes spearheaded democratic resistance in a radicalised city, collaborating with the most dazzling young thinker of the day, Coleridge, from 1796 editor of the anti-Pitt and pro-peace fortnightly, *The Watchman*.[32] Thereafter the stuffing went out of radical protest, partly because a reactionary war gradually turned into a struggle of national survival against Napo-

---

[28] Clarke (1965).
[29] Neve (1984); Greenacre (1973).
[30] Holmes (1989).
[31] Beddoes (1795, 1796a,b, 1797).
[32] These years are well discussed in Holmes (1989); Cottle (1970); Harrison (1988).

leon's brutal imperialism. Erstwhile 'friends of liberty' jumped ship and changed sides, Southey and later Coleridge beginning their long slither into 'Church and State' High Toryism. Beddoes himself grew politically quieter and perhaps demoralised, but his anti-Establishment opinions never wavered: he would always speak in a particularly truculent manner of 'us plebeians'. Ever what Dr Johnson called 'a good hater', he internalised his disappointment, and learned in those dark days to taste wormwood and gall.

Beddoes also had other fish to fry. Around 1790, he had dreamed that a new age was dawning, not just in politics but in science. Exhilarating breakthroughs in gas chemistry, above all the discovery of oxygen, would transform science and also produce astonishing medical advances. After withdrawing to Bristol, these dreams centred upon establishing a new medico-scientific site, the Pneumatic Institute, combining a laboratory with clinical services, to experiment into the curative potential of gases. Through the 1790s, Beddoes nursed these plans with phenomenal energy, cooperating with James Watt, who designed apparatus and provided funds. In works like *Considerations on the Medicinal Use of Factitious Airs* (1794), Beddoes announced his researches into a range of diseases, showing an unusual combination of clinical findings, physiological expertise, an experimental itch and a boldness in embracing new speculations.[33]

Eventually, thanks to Thomas Wedgwood's generosity, the Pneumatic Institute opened in 1799. It gave Beddoes an opportunity to intensify his gas researches; brilliant experiments with nitrous oxide (laughing gas) were conducted by Beddoes's dazzling teenage *protégé*, Humphry Davy. But it did not prove a medical success: no clinical benefits followed, funds failed, the philosophic revellers, drunk on air, were taunted in the anti-Jacobin press, and in 1801, perhaps alarmed for his reputation, Davy scuttled off to London. Retrenchment, reorganisation and rationalisation were needed; by 1802, the Institute had been revamped into the Preventive Institution, scaled-down in its scientific programme but with a revitalised mission of safeguarding the health of the poor through education and out-patient facilities.[34] Once again, however, Beddoes's hopes were dashed.

The 1800s proved wretched and demanded private virtues. Political reaction was in the saddle and radical outrage upstaged. Beddoes's crusades for science had come to little. His marriage plunged into crisis. One senses a man growing bewildered and embittered, busy but losing his bearings. Perhaps,

[33] Stansfield and Stansfield (1986); Smith (1982).
[34] Beddoes (1804).

however, it was these disappointments that led him to reflect most fruitfully upon the human comedy, especially as manifest in that bizarre blend of self-importance and suffering exhibited by the posh patients he daily attended. Certainly his most substantial writings in the new century – the remarkable thousand-page *Hygëia: or Essays Moral and Medical, on the Causes Affecting the Personal State of our Middling and Affluent Classes*, which came out in eleven monthly instalments in 1802 and 1803, and the literary experiment, the *Manual of Health: or, the Invalid Conducted Safely Through the Seasons* (1806) – engage with the foibles of his fellow men with a depth, if also a despair, absent before.[35] Ostensibly health guides for the educated, these works may be read as a hybrid between medical autobiography and scathing soap opera, a death's jest book in which Beddoes formulated, almost in a free-associating manner, his final reflections upon the theatre (or sociology) of sickness in the modern commercial and industrial order, a salmagundi of observations upon man, society, language, and literature, written in styles by turns experimental, facetious, hectoring, self-indulgent, and anything but typical of health-care manuals. Pathology seems to have been leading the doctor into new ways of observing human beings and writing about them.

His own health long deteriorating, Beddoes died, almost forgotten, on Christmas Eve, 1808, just forty-eight. Although he expired prematurely, he had outlived his time. His radical, liberal faith in the marriage of science and revolutionary principles had long been overtaken by events. For Romantic idealism had superseded Enlightenment populism; poetry had migrated to the Lakes; and the chemical crown had passed to the social climber, Humphry Davy, by then no longer at the Pneumatic Institute but at the Royal Institution in chic Piccadilly.[36] In Bristol, a new genteel intelligentsia was grouping, soon to adopt a culture of liberal Anglicanism and Peelite politics.[37]

Drama and diagnosis

A scientific experimenter, Beddoes enjoyed experiments in writing, *essais* in the Montaignian sense. As a physician, he experimented in the most literal manner, inflating the lungs of his charity patients with all manner of gases.

---

[35] Beddoes (1802, 1806).
[36] Neve (1980).
[37] Neve (1984).

But he also experimented more figuratively, filling their noddles with his intellectual vapours. Disdaining his genteel clients as *malades imaginaires* and know-all busybodies, and despairing of his indigent patients (ignorant, superstitious and clotheared), Beddoes was driven, almost in a Chekhovian mode, to translate his clinical encounters into cathartic dramas. Into these he would inscribe versions of himself in the guise of the doctor, sometimes as hero, sometimes as whipping-boy, buffoon or jester, testing out the varied inflections of the healer's art, its hopes and hubris. The habit of projection first assumed public form in his early poem, *Alexander's Expedition down the Hydaspes & the Indus to the Indian Ocean* (1792). Beddoes used Alexander's incursion as a vehicle for denunciation of British imperialism in India (recently focused by the Warren Hastings scandal), where, thanks to monarchical militarism, 'plague and rapine' were everywhere rampant:

> But earth's fond hopes, how blasted in their bloom!
> How feels a world convulsed by early doom!
> What mingling sounds of woe and outrage rise!
> How wide the eddying dust of ruin flies![38]

'Degraded by oppression', the conquered Indians – Beddoes's metonym for the downtrodden world-wide – deserved 'pity'.[39] Their plight would remain beyond relief, 'Till the last hour of their merciless tyrants from Europe shall arrive', evil multiplying till we 'banish slavery and despotism of every species from the face of the earth'.[40] But alongside radical propaganda, Beddoes also created a rather novel and clearly autobiographically influenced reading of the Macedonian. Traditionally a hero, Alexander had become, in much eighteenth-century thinking, a power-crazed tyrant, literally drunk with power, epitomising unregulated military despotism.[41] Beddoes switched the image, and projected him as a fervent, questing Romantic, driven by a Byronic craving for experience to burst the bounds of a stifling world. 'His mind', he judged, 'was discriminated by exquisite sensibility. By whatever object they were touched, the springs of his Nature bent deeply inwards; but they immediately rebounded with equal energy into action. Hence one may explain his passionate excesses: that independence of mind that would not blindly submit even to an Aristotle'. Alexander was the great overreacher,

---

[38] Beddoes (1792*a*), vi.
[39] Beddoes (1792*a*), 13.
[40] Ibid.
[41] Braudy (1987). While thinking of Alexander as a megalomaniac, Beddoes also self-identified with him as a fellow young man of genius. Alexander's drinking problem is assessed in O'Brien (1992).

advancing 'magnificent designs which exceeded the comprehension of his age. Thus his genius was doubtless great'.[42] The promethian self-identification is transparent.

Insofar as Beddoes empathised with an Alexander whose mission was to impose enlightenment upon a fractious world, it was a projection Beddoes retained all his life, though he rang the changes on its patterns. Sometimes he inserted himself into his fantasy fictions as a benevolent, all-wise doctor of humanity. In *The History of Isaac Jenkins, and of the Sickness of Sarah his Wife, and their Three Children* (1792), a Shropshire surgeon, Mr Langford – remember Beddoes was himself a Shropshire lad – reconquers a family from drink, despair and quackery. The moral tract relates a simple tale. It is the bleak midwinter of 1783. Isaac Jenkins, a labourer, Sarah, his wife, and their three youngsters, live in Titterstone, a Shropshire village known to Beddoes since childhood. Illness overtakes the family. Unable to afford the doctor, they buy potions off the local quack. When they can't pay him any more, the huckster naturally stops calling ('your quack doctors', asides Beddoes, 'care not a farthing whether they cure or kill, all they want is to fleece those that know no better').[43]

Luckily, Surgeon Langford (a.k.a. Beddoes) rides past one day *en route* to attend the parson, injured (typical touch!) while out shooting. Langford gets talking to Sarah, hears the children are sick, treats them free, and listens to her tale of woe. Isaac has taken to drink (though this lapse stems, inter-polates the preachy Beddoes, not from inherent depravity, but from a terrible accident that had befallen the family: 'I believe for my part that the poor are well disposed, and do wrong oftener for want of knowing better than from wickedness').[44]

Langford has a heart-to-heart with Isaac, prescribes medicines, and shames him into resuming his long-neglected family responsibilities. A loan from the surgeon enables the family to repay the publican. Following doc-tor's orders, Isaac recovers – unlike his odious master, who (another Beddoe-sian class barb) comes to a disgusting end, over-indulgence leading to death from dropsy. All ends happily, with Isaac learning the difference between 'plenty with sobriety and a light and cheerful heart, and beggary with drunk-enness and discontent'.[45] An 'epilogue' for the 'refined reader' urges that,

---

[42] Beddoes (1792*a*), 13.
[43] Beddoes (1792*b*), 5; of the quack, the narrator notes, 'two out of three died of those that he doc-tored, though he was a conjuror beside'.
[44] Ibid., 37.
[45] Ibid., 40.

following Locke's psychology, the poor should not be stigmatised as born reprobates, but viewed as victims of circumstances, capable of coming good, given proper education and instruction.[46]

A similar upright doctor character appears to 'rescue' the poor in Beddoes's brochure about the Pneumatic Institute. In the *Rules of the Medical Institution, for the Benefit of the Sick and Drooping Poor* (1804), Beddoes set out his preventive programme for saving the poor from disease, from themselves, from ignorant friends and from kitchen physic. Unlike the success story in *Isaac Jenkins* however, here, ten years later, the heroic doctor is much abused. Patients fail to turn up, fail to listen, fail to take their medicine. 'Confidence is placed in ignorant neighbours' rather than doctors, Beddoes bemoaned, 'because they speak to the sick in their own language, and pretend to give some sort of reason for what they recommend'. Distant doctors left the patient with 'no conception of the turns his disorder must take before it can go off ':

> he is cast down without cause. He goes to another doctor, and from him to a quack. Nay, in some parts, poor sick people cannot be persuaded to go to any body but to the most silly old women. Here, in Bristol, I often find sad mischief done by some improper medicine, and it turns out nine times out of ten that the medicine has been advised by some ignorant neighbour.[47]

A guilt-ridden rescuer, Beddoes here pillories himself, like many of the anguished do-gooders in the Godwinian and sentimental novels of the day. For the delinquencies of the poor, Beddoes tended to blame his ilk: 'all has hitherto been conducted in a style of authority. It has been too much mere dumb shew between doctor and patient'.[48] Yet in such professional incrimination, a double process is at work, for self-accusation is self-exoneration. *Rules of the Medical Institution* fuses apologia and self-flagellation, humility and superiority.

The noble doctor, a prophet despised in his own country, also plays a lead part in *Hygëia* (1802) and the *Manual of Health* (1806), Beddoes's Shavian tragi-comedies of health care for the rich. Here, however, he is a butt, impotent in his battles against perverse genteel stupidity. 'What is good against the head-ache, Doctor?', inquires a typical patient in one of Beddoes's little pantomimes. 'Health, Madam', he replies:

---

[46] Ibid., 43.
[47] Beddoes (1804), 95.
[48] Ibid., 95. For fictional parallels of self-laceration, see Kelly (1976).

'Well, if you feel no interest about an old woman like me – Mari-anne there, you perceive, has been hacking all the afternoon. Do tell her of some little thing, that is good against a cough.' 'Health, Madam.' 'But are you resolved not to give a more satisfactory answer? In that case, I shall take the liberty of guessing why.' 'Poh! Mrs W—', cried a grave person in spectacles, from behind a full hand of cards – 'you should know that it is the trick of these gentle-men NEVER TO SPEAK PLAIN, as some great man says. And if they will not in a tête-a-tête, can you expect it before company?' 'I am not conscious of having uttered any enigma. I am sorry for the young lady. But I must still answer – health, Sir, health, Madam'.[49]

In this skit, the physician plays the wise fool, almost silent, his words unheeded. His recreation of the farce – one surely experienced endlessly on his rounds to the rich – doubtless sublimates his rage. Beddoes concludes by noting 'the lady it will appear from the preceding little fragment, belongs to the corps, from which I am ambitious to enlist recruits';[50] but this is not to be taken at face value, since, by this stage, it is unthinkable that Beddoes truly expected to convert the fashionably foolish, even though he attended them, and meant them to buy his books. Beddoes's attachment to the cause of health is not in question: he went so far as to set up lecture courses in hygiene and physiology. But his contempt for his more pretentious patients (and readers) is equally clear: 'The author is certainly accustomed to see invalids, for whom it would be happy if their whole mass of ideas – provided those were included that relate to the means of their recovery – could be abolished'.[51] For this avowed revolutionary, it would be a small leap from the abolition of his patients' mental clutter to the abolition of the patients themselves. Beddoes obviously found much about the sick, and the sick society, hard to stomach.

'Take physic, physic':[52] auto-diagnosis

Beddoes thus sketched portraits of the physician as hero, tragic or comic, slighted and snubbed (and hence as perfect fool). Such self-dramatisations

---

[49] Beddoes (1802), 1 ii 14 and 15.
[50] Ibid., 1 ii 15. For the tradition of playing the fool, see Billington (1984); Kaiser (1964); Welsford (1935).
[51] Ibid., 1 i 53.
[52] The motto is used as the epigraph for *Hygëia*. It echoes *King Lear*.

presumably offered oblique gratification, allowing him to strut as an exemplary figure of sorts. But what is particularly interesting is that Beddoes seems to have offered not just savage pathographies of others but autodiagnostics too. In his writings the doctor/character also crops up questioned, undercut by his own analyses, subject to self-critique. Using parody, exaggeration and irony, Beddoes thereby presents himself as part of the problem rather than its solution; or rather both at the same time.

In some measure, this hypochondriac effect is enhanced through Beddoes's obsessional lashing of his own profession. He was a merciless critic of medical men. Quacks he abhorred, but he insisted that, greedy for guineas and trafficking in the 'sick trade', regulars were no better; in fact doctors and quacks were two peas in a pod — medical Jesuits and medical Jansenists, he tellingly suggested:

> Without going a hundred miles from Clifton, Bristol and Bath, you may meet with practitioners, whose genius has transported them at a single bound from the side of the mortar to the bedside; and who go about distributing their *poudres de succession* through town and country, with as much professional gravity, as if they had gone through the longest course of study, and stood the severest trials of skill. As to consumption-doctresses, cancer-curers, mechanics professing to treat divers disorders, and particularly those of the female sex, there have arisen within my short memory, several, in whose behalf to speak with the cricket players, one might safely challenge all England; nay, in this favoured district, do we not behold the splendid seats of solemn, regular, respected quackery, methodized, as you shall hear, if you do not know it, according to the most approved forms of a foreign merchant's counting house.[53]

Beddoes thus lumped together all practitioners under the category of commercial traders. No accident, maybe, that the centre of the circle was the heartland of his own practice.

Beddoes naturally liked to personate off as a cut above all that: as a high-minded friend of science and humanity, indifferent to lucre.[54] And doubtless he was. But it was not so simple. After all, he had opted to practise in an extremely well-to-do residential area; he gained and maintained a fashionable clientele, to whose foibles he evidently pandered (in many of his mini-dramas we can hear the enforced grovelling). Beddoes himself started poor:

[53] Beddoes (1808), 9.
[54] See Porter (1991a), ch. 8 *passim*.

lacking inherited wealth, he married into a family with high expectations, he had young children, and every spare penny was needed for funding his experiments and the Pneumatic Institute. For mixed motives, Beddoes was no doubt as mercenary as the money-mad physicians he accused. It is remarkable that his official biographer did not merely note that he had 'been charged with avarice and with an eager and undistinguishing rapacity for fees', but accepted these charges were 'partially founded'.[55] Beddoes's diatribes against the doctors thus support the view that, privately at least, he acknowledged his own complicity in the practices he condemned.

Self-doubt and self-diagnosis loom large in another of his preoccupations. Beddoes believed modern times had brought upon themselves an epidemic of self-inflicted troubles, diseases of civilisation akin to the condition George Cheyne had identified three quarters of a century earlier as the 'English malady'.[56] The nation had grown rich and hence refined, and over-refinement created a kind of indolence ('oppressive languor and sickliness'), attended by killing time and ruining constitutions in the 'carpeted, stuccoed, and stoved sitting room'.[57] Leisure brought other evils in its train – above all, paradoxically, the business of doing nothing, the hectic hurry of modern vacancy.

England had also been engulfed in the vortex of fashion and opinion in a newly-polite Addisonian culture. Fashion demanded that anyone who was anyone be *au fait* with the latest ideas, plays and pamphlets. Novelty was all the rage, change the driving demon. Traditional, well-tried ways of living, Beddoes judged, were yielding to smart silliness. The tea table, the circulating library, the premature cramming of children with book-learning and excessive cultivation of sensibilities – all these were producing enfeeblement. Beddoes – he was not alone in this – believed that Addisonian culture was literally bad for people's health. Minds were racing, nerves growing frayed, and all for what? – 'the despotism of fancy'. 'What elicits . . . sparks of emulation from the eyes, and induces agitations that equally disturb the rest, of thousands of striplings and of damsels, panting for celebrity in Britain?' teased Beddoes – 'why, may be, a nosegay of artificial plumes, or a well-stiffened collar'.[58] It was impossible to keep up. *'Did you see the papers today? Have you read the new play – the new poem – the new pamphlet – the last novel?'*, Beddoes reproved, was all one heard nowadays. 'You cannot

[55] Stock (1811), 397.
[56] Porter (1990), 'Introduction'; Porter (1991*b*).
[57] Beddoes (1802), 2 v 60.
[58] Ibid., 1 i 8.

creditably frequent intelligent company, without being prepared to answer these questions, and the progeny that springs from them'. The consequence? 'You must needs hang your heavy head, and roll your bloodshot eyes over thousands of pages weekly. Of their contents at the week's end, you will know about as much as of a district, through which you have been whirled night and day in the mail-coach'. Sick headaches and enervation were the symptoms of this tyranny of 'quick desultory reading': 'What wonder then that we should hear complaints against the age as wanting energy of feeling and compass of mind?'[59]

Fast living, idle business, and the itch for novelty – the modern buzz words were *'as soon as possible'* and *'as fast as possible!'*[60] – reinforced emulation and competition, the urge to sparkle in the world's eye, seeing and being seen. 'An universal interest must be taken in the condition of those about us', for everybody wanted to dazzle 'strangers by the splendour of an equipage or by the lights of the understanding'. We want 'every thing about us' – note the first person – 'cloaths, tables, chairs, pictures, statues – all exquisite in their kind'; the only item neglected in this sumptuous parade? – health.

Polite culture in commercial civilisation thus jeopardised the well-being of the affluent and aspirant. In smart circles, Beddoes chid, the *comme il faut* was to pass oneself off as witty, brimful of news, tasteful opinions, and 'smart tripping chat' about faces and places. The better sort were bingeing on booklets as once they had on beefsteaks – and the inevitable *sequela* of such 'cramming the head with dose after dose of heterogeneous ideas, before the first have had time to settle' was intellectual indigestion.[61] Days better given over to the farm or the field were being wasted on the sofa, eyes glued to duodecimos. And what were folks the better for all this diet of instant opinions and *ersatz* stimulus? Beddoes took a very dim view of this addling of the elite's brains by light reading, especially those terrible innovations, novels, which 'render the sensibility still more diseased'. Of all popular writing 'NOVELS undoubtedly are the sort most injurious . . . They increase indolence, the imaginary world indisposing those, who inhabit it in thought, to go abroad into the real'.[62] Fiction fostered fantasy: 'The common love-stories . . . relax soul and body at once' – and, Beddoes implied, though in

[59] Ibid., 3 ix 163. *'Hypocrite auteur, mon semblable, mon frère'*, one is tempted to respond.
[60] Ibid., 1 ii 65.
[61] Ibid., 1 iii 32. For background, see Langford (1989, 1991); Sekora (1977).
[62] Ibid., 1 iii 77.

the most mealy-mouthed terms, they tempted teenagers into mastur-bation.[63]

Overall, Beddoes concluded, 'it seems impossible to live in the fret and fume of fashion without direct injury'.[64] Yet that was precisely where he was living, and he must have recognised himself as contributing to the problem, and, presumably too, as being one of its victims. A prolific writer, a tireless contributor to reviews and journals, both medical and literary, a passionate campaigner for education, Beddoes added far more than his fair share to the literary ephemera of his times. Was not the *Manual for Health* consumed by fashionable ladies lolling on *chaises longues*, seeking diversion between novels? In this light, Beddoes might almost be envisaged as a medical por-nographer, seducing readers into doing precisely those things – consuming polite *belles lettres* – against which he was admonishing them.

Did Beddoes see himself as a medical-culture pusher and junkie?[65] After all, if anyone's health was cracking up it was his own. From his thirties, he grew tubby and short-winded. It is likely that self-experimentation with nitrous oxide, other gases and narcotics undermined his constitution. He probably suffered from overwork and from precisely the kind of hectic life-style leading to *ennui* and a 'want of relish'[66] he so avidly censured in his age. When he wrote of folks spicing up their dreary days with 'provocatives' like alcohol and narcotics that merely 'render the succeeding listlessness more intolerable, and another forced orgasm more indispensable', did he have an eye upon himself?[67]

And what was the relation between his health and his erotic identity? Beddoes was remarkably preoccupied with the evils of self-abuse. In various writings, he characterised onanism as an evil often contracted in youth and encouraged by genteel sedentary habits and by light reading that increased 'indolence, the imaginary world indisposing those, who inhabit it in thought, to go abroad into the real'.[68] Once imaginations were thus inflamed, the results were dire, for self-abuse led to debility and disease.

---

[63] Ibid., 1 iii 78. On masturbation see see Boucé (1985); Hare (1962); MacDonald (1967); Stengers and Van Neck (1984); Jordanova (1987). Note that in the eighteenth century the term 'relax' had negative medical connotations ('cause lassitude').

[64] Ibid., 1 ii 70.

[65] Porter (1994), explores the genre of writing that warns readers of the perils of reading.

[66] Beddoes (1802), 1 ii 62.

[67] Ibid., 1 ii 70. 'Orgasm' here does not have a narrowly sexual meaning. See also 1 ii 65; 1 i 62, and for the wider problem of alcohol and addiction, see Porter (1989c), and Trotter (1807). For Beddoes's declining health, see Stansfield (1984), 237.

[68] Beddoes (1802), 2 vi 72.

Onanism was thus another disorder caused, or exacerbated, by the suggestive, seductive power of language and imagination. Masturbation, Beddoes believed, had been the cause of Jonathan Swift's premature senility, with its 'loss of associative power' and attendant 'nervous complaints', culminating in the 'madness of misanthropy'.[69] How far did he associate writing with masturbating, seeing both as pathological? Did he link himself with Swift?

From around his fortieth year, Beddoes clearly felt in decline. We know little about his sexual life and habits, beyond that he wed in his mid-thirties and that within a few years his marriage was failing (his wife temporarily left him, flinging herself upon his best friend, Giddy). Whether Beddoes looked to some sexual vice in himself to explain his decline cannot be ascertained. But there may be one clue. In a last letter to Davy just before his death, Beddoes depicted himself in failure as 'like one who has scattered abroad the Avena fatua of knowledge from which neither brand, nor blossom nor fruit has resulted'.[70] The image of intellectual wild oats barrenly scattered strikingly echoes the sin of Onan, spilling his seed fruitlessly upon the ground. Onanism has recently been characterised as the typical disorder of heightened imaginative stimulus associated with Enlightenment habits of private writing and reading. This would broadly seem to fit Beddoes. Is this the key to 'Take Physic, Physic'? It seems a bitter pill.

Conclusion

Uniting writer and clinician, Beddoes was to recognise that illness was not simply an adventitious foe. It was more like a familiar or a double, integral to sufferers' lives: wealth-specific, class-specific, gender-specific, *mores*-specific. The attentive doctor could thus be a diagnostician of social pathology in the body politic and in social circles, and maybe in himself.

The archetypal auto-diagnosis is Freud's. Freud's self-analysis appears to have left him strong. The contrary was perhaps true of Beddoes. Brooding over political failure in 'cannibal times', Beddoes's lashing of his age was surely a self-flagellation, using the scourge of self-deflating irony.

Beddoes hoped that, as *philosophe-médecin*, educator and writer, he could promote solutions to ill health. He was also sufficiently self-aware that he and his medical peers might be pathogenic. After all, his profession had to

---

[69] Ibid., 1 i 45, 1 i 48. See further discussion in Porter (1991c, 1993).
[70] Stansfield (1984), 249.

shoulder responsibility for the perversions of practice, for it was they who were profiteering from such obscene temples to Mammon as fashionable spas.[71] After all, the spirit of the Enlightenment was for putting medicine in lay hands.[72] After all, Beddoes was himself a compulsive writer of tomes targeted at the public's tired eyes and reeling brains. If books were addictive, could Beddoes himself — writing books against bookishness in a manner compounding the paradoxes of Robert Burton and Laurence Sterne — escape censure? ('Will this be good for your worship's eyes?', questioned Sterne.) Beddoes was not unaware of the ironies of his own stance.[73]

The cause of health hinged upon education. Would it work? – for everyone wanted to teach, no one to learn. It was a 'foible', he believed, of 'instructors' and 'us authors', to 'take much greater pleasure in giving than in following good advice. Both would rather find fault in others, than amend their own'.[74] How willing was Beddoes to heed his own advice, take his own medicine? Beddoes devoted 200 pages of that work of 'moral materia medica', *Hygëia*, to savaging public schools for boys and boarding-schools for girls.[75] Yet he never wavered from his Locke- and Hartley-derived vision of mankind as essentially educable ('there is much less disparity in the power of apprehension of different ranks, than the superior members of society are willing to flatter themselves').[76] What was crucial was the right method: teaching 'can very seldom be done by threats or by commands'.[77] There were grounds for hope, and certainly scope for improvement.

But Beddoes also grew exasperated with sickness on the personal plane. Illness was woven into the thick texture of living. It was a mode of self-experience; it could cement group identity and form a medium of social exchange.[78] Long before Talcott Parsons's coinage, Beddoes was well aware of the equivocations of the 'sick role';[79] it was a part the patient could play; not least, it was a cultural *Träger* with a life of its own, victimising the spirits

---

[71] See Porter (1991a), ch. 7.
[72] Porter (1992b).
[73] Sterne (1967), 265. See also Porter (1984). Stock noted Beddoes's admiration for Sterne, and spoke of the 'Shandyism in his philosophy': 'Sterne was one of his favourite English authors, and he had, at one period of his life, collected some materials with a view to give to the world an Essay on his life and character'. Stock (1811), 347, 410.
[74] Beddoes (1802), 1 iii 15.
[75] Ibid., 1 iii 59.
[76] Ibid., 2 v 16.
[77] Ibid., 2 v 16. For the psychological background see Yolton (1956); Young (1973).
[78] For scholarship discussing these dimensions, see Brody (1987); Kleinman (1980, 1986, 1988); Porter and Porter (1988).
[79] Parsons (1951).

and penalising the bodies of those it scripted. Beddoes was intrigued by sickness as fetish, as the object of ritual and intricate belief-systems, as a dialect of social intercourse, as the hermeneutic key to habits; sickness, in other words, as cult, an emotionally entangled language of life which, however irrational, articulated moods and emotion, and radiated streams of meaning. Beddoes fumed at the absurd health-lore practices of poor people and the self-indulgent health antics of rich folks who should have known better. But, like a Christian missionary meticulously recording obscene pagan ceremonials, Beddoes became enwrapped in all those fictions and fantasies, affectations, silly clichés and stock *repartée* that took the terror out of disease by enabling them to be spoken; and he documented this exotic oral culture in remarkable detail, with the ear of a fine doorstep sociologist, a sympathetic novelist, participant observer, and a satirical caricaturist rolled into one. And what he heard he wrote down, while pondering whether, like other projectors, he might have fallen a 'dupe to self deception'.[80]

Two conclusions thus suggest themselves from this case-study as answers to my opening question regarding the specificity of medical biography. The genre spotlights a rich and ambivalent category – sickness – that can serve to open windows onto individuals and groups, precisely because of its capacity to bridge the chasm between the objective and the subjective, the outer and the inner, the public and the private. The doctor thus becomes a kind of guide and interpreter of foreign parts. Secondly, the case of Beddoes parades the image of doctor as both observer and participant. That is surely the doctor's distinctive vocation, the capacity to see himself in others and others in himself. This dialectic, this double vision, this duplicity is surely also the biographer's duty and privilege.

Bibliography

Abraham, J.J. (1933) *Lettsom: His Life, Times, Friends and Descendants.* London: Heinemann.

Andrews, A.F. (1896) *The Doctor in History, Literature, Folk-Lore.* London: Hull & Andrews.

Anecdotes Medical, Chemical and Chirurgical; Collected, Arranged and Transmuted by An Adept (1816), 2 vols. London: Callow.

Beddoes, T. (1787) *A Memorial Concerning the State of the Bodleian Library, and the Conduct of the Principal Librarian. Addressed to the Curators of that Library, by the Chemical Reader.* Oxford: University Press.

---

[80] Beddoes (1802) 1 ii 14 and 15. Beddoes's emphasis upon the positive importance of health squares well with the discussion in Foucault (1976) except, of course, that Beddoes claimed the bourgeoisie neglected their health whereas Foucault's argument is that they seriously cultivated it.

Beddoes, T. (ed.) (1789), L. Spallanzani, *Dissertations Relative to the Natural History of Animals and Vegetables*, 2nd edition. London: J. Murray.

Beddoes, T. (1792*a*) *Alexander's Expedition down the Hydaspes & the Indus to the Indian Ocean*. Madeley: J. Edmunds.

Beddoes, T. (1792*b*) *The History of Isaac Jenkins, and of the Sickness of Sarah his Wife, and their Three Children*. Madeley: J. Edmunds.

Beddoes, T. (1792*c*) *Extract of a Letter on Early Instruction, Particularly that of the Poor*. Madeley: J. Edmunds.

Beddoes, T. (1795) *Where would be the Harm of a Speedy Peace?* Bristol: N. Biggs.

Beddoes, T. (1796*a*) *An Essay on the Public Merits of Mr Pitt*. London: J. Johnson.

Beddoes, T. (1796*b*) *A Letter to the Right Hon William Pitt, on the Means of Relieving the Present Scarcity, and Preventing the Diseases that Arise from Meagre Food*. London: J. Johnson.

Beddoes, T. (1797) *Alternatives Compared: or, What Shall the Rich Do to be Safe?* London: J. Debrett.

Beddoes, T. (1802) *Hygëia: or Essays Moral and Medical, on the Causes Affecting the Personal State of our Middling and Affluent Classes*, 3 vols. Bristol: J. Mills.

Beddoes, T. (1804) *Rules of the Medical Institution, for the Benefit of the Sick and Drooping Poor; with an Explanation of its Peculiar Design and Various Necessary Instructions*. Bristol: J. Mills.

Beddoes, T. (1806) *Manual of Health: or, the Invalid Conducted Safely Through the Seasons*. London: Johnson.

Beddoes, T. (1808) *A Letter to the Right Honourable Sir Joseph Banks . . . on the Causes and Removal of the Prevailing Discontents, Imperfections, and Abuses, in Medicine*. London: Richard Phillips.

Benjamin, M. (1990) Medicine, morality and the politics of Berkeley's tar-water. In Cunningham and French (1990), 165–93

Billington, S. (1984) *The Social History of the Fool*. Brighton: Harvester Press.

Black, J. and Gregory, J. (eds) (1991) *Culture, Politics and Society in Britain 1660–1800*. Manchester: Manchester University Press.

Booth, C. (1987) *Doctors in Science and Society: Essays of a Clinical Scientist*. London: British Medical Journal.

Boucé, P.-G. (1985) Les jeux interdits de l'imaginaire: onanisme et culpabilisation sexuelle au XVIIIe siècle. In Céard (1985), 223–43

Braudy, L. (1987) *The Frenzy of Renown*. Oxford: Oxford University Press.

Brieger, G. (1980) History of medicine. In Durbin (1980), 121–96.

Brieger, G. (1993) Historiography of medicine. In Bynum and Porter (1993), 24–44.

Brock, C.H. (1983) *William Hunter, 1718–1783*. Glasgow: Glasgow University Press.

Brody, H. (1987) *Stories of Sickness*. New Haven: Yale University Press.

Brown, J.T. (1903) *Dr John Brown: A Biography and a Criticism* (ed. W.B. Dunlop). London: Black.

Burke, P. and Porter, R. (eds.) (1991) *Language, Self and Society: The Social History of Language*. Cambridge: Polity Press.

Butler, M. (1981) *Romantics, Rebels and Reactionaries: English Literature and its Background 1760–1830*. Oxford: Oxford University Press.

Bynum, W.F. (1980) Health, disease and medical care. In Rousseau and Porter, 211–54.

Bynum, W.F. and Porter R. (eds) (1985) *William Hunter and the Eighteenth Century Medical World*. Cambridge: Cambridge University Press.

Bynum, W.F. and Porter R. (eds) (1993) *The Routledge Companion Encyclopaedia of the History of Medicine*. London: Routledge.

Céard, J. (ed) (1985) *La Folie et le Corps*. Paris: Presses de l'Ecole Normale Supérieure.

Christie, J. and Shuttleworth, S. (eds) (1989) *Nature Transfigured*. Manchester: Manchester University Press.

Clark, R. (1980) *Freud: The Man and the Cause*. New York: Random Press and London: Jonathan Cape/Weidenfield & Nicolson.

Clarke, D. (1965) *The Ingenious Mr Edgeworth*. London: Oldbourne.

Cockshutt, A.O.J. (1984) *The Art of Autobiography in Nineteenth and Twentieth Century England*. New Haven: Yale University Press.

Cone, C.B. (1968) *The English Jacobins: Reformers in Late 18th Century England*. New York: Scribner.

Corner, B.C. and Booth, C. (eds) (1971) *Chain of Friendship: Selected Letters of Dr John Fothergill*. Cambridge, MA: Harvard University Press.

Corsi, P. & Weindling, P. (eds.) (1983) *Information Sources in the History of Science and Medicine*. London: Butterworth.

Cottle, J. (1970) *Reminiscences of Samuel Taylor Coleridge and Robert Southey*. Highgate: Lime Tree Bower Press [1st edition (1847) London: Houlston & Stoneman].

Cousins, N. (1982) *The Physician in Literature*. Philadelphia: W.B. Saunders.

Cunningham, A. & French, R. (eds) (1990) *The Medical Enlightenment of the Eighteenth Century*. Cambridge: Cambridge University Press.

DeMause, L. (1975) *The New Psychohistory*. New York: Psychohistory Press.

Duden, B. (1987) *Geschichte unter der Haut*. Stuttgart: Klett, Cotta. Translated by Thomas Dunlap (1991) as *The Woman Beneath the Skin. A Doctor's Patients in Eighteenth-Century Germany*. Cambridge, MA: Harvard University Press.

Durbin, P.T. (ed.) (1980) *A Guide to the Culture of Science, Technology and Medicine*. New York: Free Press.

Dusseau, J.L. (1979) The Great Cham as a medical biographer. *Pharos*, **42**, 10–17

Flynn, C.H. (1991) Running out of matter: the body exercised in eighteenth-century fiction. In Rousseau (1991*c*), 147–85

Forgan, S. *et al.* (eds) (1980) *Science and the Sons of Genius: Studies on Humphry Davy*. London: Science Reviews.

Foucault, M. (1976) *Histoire de la sexualité*: Vol.1. *La volonté de savoir*. Paris: Gallimard. Translated by Robert Hurley (1978) as *The History of Sexuality: Introduction*. London: Allen Lane.

Fox, R.H. (1919) *Dr John Fothergill and His Friends: Chapters in Eighteenth Century Life*. London: Macmillan & Co.

Gabbard, K. and Gabbard, G.O. (1987) *Psychiatry and the Cinema*. Chicago: University of Chicago Press.

Gay, P. (1985) *Freud for Historians*. New York and Oxford: Oxford University Press.

Goodwin, W. (1979) *The Friends of Liberty: The English Democratic Movement in the Age of the French Revolution*. London: Hutchinson.

Greenacre, F. (1973) *The Bristol School of Artists: Francis Danby and Painting in Bristol 1810–1840* (catalogue). Bristol: the Gallery.

Greenblatt, S. (1980) *Renaissance Self-Fashioning: From More to Shakespeare*. Chicago: University of Chicago Press.

Halévy, E. (1924) *The Growth of Philosophical Radicalism*. London: Faber.

Hankins, T.L. (1979) In defence of biography: the use of biography in the history of science. *History of Science*, 17, 1–16.

Hare, E.H. (1962) Masturbatory insanity: the history of an idea. *Journal of Mental Science*, 108, 1–25.

Harrison, M. (1988) *Crowds and History: Mass Phenomena in English Towns, 1790–1835*. Cambridge: Cambridge University Press.

Heathcote, R. (1786) *Sylva; or, The Wood: Being a Collection of Anecdotes, Dissertations, Characters, Apophthegms, Original Letters, Bons Mots, and Other Little Things*. London: T. Payne & Son.

Heberden, E. (1990) *William Heberden 1710–1801: Physician of the Age of Reason*. London: Royal Society of Medicine.

Holmes, F.L. (1981) The fine structure of scientific creativity. *History of Science*, 19, 60–70

Holmes, R. (1989) *Coleridge: Early Visions*. London: Hodder & Stoughton.

Jeaffreson, J.C. (n.d.) *A Book About Doctors*. London: Hurst & Blackett.

Jordanova, L.J. (1987) The popularisation of medicine: Tissot on onanism, *Textual Practice*, 1, 68–80.

Kaiser, W. (1964) *Praisers of Folly*. London: Victor Gollancz.

Kelly, G. (1976) *The English Jacobin Novel, 1780–1805*. Oxford: Oxford University Press.

King-Hele, D. (1977) *Doctor of Revolution: The Life and Genius of Erasmus Darwin*. London: Faber.

Kingsbury, M. (1987) Congenial associates: the biographical essays of William Osler. *Biography*, 10, 225–40.

Kleinman, A. (1980) *Patients and Healers in the Context of Culture: An Exploration of the Borderline between Anthropology, Medicine, and Psychiatry*. Berkeley: University of California Press.

Kleinman, A. (1986) *Social Origins of Distress and Disease: Depression, Neurasthenia, and Pain in Modern China*. New Haven: Yale University Press.

Kleinman, A. (1988) *The Illness Narratives: Suffering, Healing and the Human Condition*. New York: Basic Books.

Langford, P. (1989) *A Polite and Commercial People: England 1727–1783*. Oxford: Oxford University Press.

Langford, P. (1991) *Public Life and the Propertied Englishman 1689–1798*. Oxford: Clarendon Press.

Lawrence, C.J. (1984) Medicine as culture: Edinburgh and the Scottish Enlightenment. University of London, Ph.D. thesis.

Leader, Z. (1991) *Writer's Block*. Baltimore: The Johns Hopkins University Press.

Levine, G. (ed.) (1987) *One Culture: Essays in Science and Literature*. Madison: University of Wisconsin Press.

Loewenberg, P. (1983) *Decoding the Past: The Psychohistorical Approach*. New York: Alfred Knopf.

MacDonald, R.H. (1967) The frightful consequences of onanism. *Journal of the History of Ideas*, 28, 423–41.

McHenry, L.C. Jr (1959) Dr Samuel Johnson's medical biographies. *Journal of the History of Medicine*, **14**, 298–310.

McKee, F. (1991) The earlier works of Bernard Mandeville, 1685–1715. Glasgow University, Ph.D. thesis.

MacNeil, M. (1987) *Under the Banner of Science. Erasmus Darwin and his Age*. Manchester: Manchester University Press.

Melling, J. and Barry, J. (eds) (1992) *Culture in History*. Exeter: Exeter Studies in History.

Micale, M. and Porter, R. (eds) (1994) *Discovering the History of Psychiatry*. New York: Oxford University Press.

Money, J. (1977) *Experience and Identity: Birmingham and the West Midlands 1760–1800*. Manchester: Manchester University Press.

Moore, J.R. (ed.) (1989) *History Humanity and Evolution*. Cambridge: Cambridge University Press.

Morris, D.B. (1991) *The Culture of Pain*. Berkeley: University of California Press.

Morton, L.T. and Moore, R.J. (1989) *A Bibliography of Medical and Biomedical Biography*. Aldershot: Scolar Press.

Musson, A.E. and Robinson, E. (1969) *Science and Technology in the Industrial Revolution*. Manchester: Manchester University Press.

Myer, V.G. (ed.) (1984), *Laurence Sterne: Riddles and Mysteries*. London and New York: Vision Press.

Neve, M.R. (1980) The young Humphry Davy: or John Tonkin's lament. In Forgan *et al* (1980), 1–33.

Neve, M.R. (1984) Natural philosophy, medicine and the culture of science in provincial England: the cases of Bristol, 1790–1850, and Bath. University of London, Ph.D. thesis.

O'Brien, J.M. (1992) *Alexander the Great. An Invisible Enemy*. London: Routledge.

Outram, D. (1978) The language of natural power: the *éloges* of Georges Cuvier and the public language of nineteenth century science. *History of Science*, **16**, 153–78.

Parsons, T. (1951) *The Social System*. London: Routledge & Kegan Paul.

Paul, C.B. (1980) *Science and Immortality: The Eloges of the Paris Academy of Science (1699–1791)*. Berkeley: University of California Press.

Pelling, M. (1983) Medicine since 1500. In Corsi and Weindling (1983), 379–407.

Peterfreund, S. (ed.) (1990) *Literature and Science: Theory and Practice*. Boston: Northeastern University Press.

Plumb, J.H. (1969) *The Death of the Past*. London: Macmillan.

Porter, D. and Porter, R. (eds.) (1993) *Doctors, Politics and Society*. Amsterdam: Rodopi.

Porter, R. (1977) *The Making of Geology: Earth Science in Britain, 1660–1815*. Cambridge: Cambridge University Press.

Porter, R. (1984) Against the spleen. In Myer (1984), 84–99.

Porter, R. (1989*a*) Erasmus Darwin: doctor of evolution? In Moore (1989), 39–69.

Porter, R. (1989*b*) 'The Whole Secret of Health': mind, body and medicine in *Tristram Shandy*. In Christie and Shuttleworth 61–84.

Porter, R. (1989*c*) Introduction. In Trotter (1989).

Porter, R. (ed) (1990) *George Cheyne: the English Malady*. London: Routledge. [1st edition 1733.]

Porter, R. (1991*a*) *Doctor of Society: Thomas Beddoes and the Sick Trade in Late Enlightenment England*. London: Routledge.

Porter, R. (1991*b*) Civilisation and disease: medical ideology in the Enlightenment. In Black and Gregory (1991), 154–83.

Porter, R. (1991*c*) 'Expressing Yourself Ill': the language of sickness in Georgian England. In Burke and Porter (1991), 276–99.

Porter, R. (1992*a*) Addicted to modernity: nervousness in the early consumer society. In Melling and Barry (1992), 180–94.

Porter, R. (1992*b*) Spreading medical enlightenment: the popularisation of medicine in Georgian England, and its paradoxes. In Porter (1992*c*), 215–31.

Porter, R. (ed) (1992*c*) *The Popularization of Medicine, 1650–1850*. London: Routledge.

Porter, R. (1994) The literature of sexual advice before 1800. In Porter and Teich (1994), 134–57.

Porter, R. and Porter D. (1988) *In Sickness and in Health. The British Experience, 1650–1850*. London: Fourth Estate.

Porter, R. and and Teich, M. (eds) (1994) *Sexual Knowledge, Sexual Science*. Cambridge: Cambridge University Press.

Roberts, M.M. and Porter, R. (eds) (1993) *Literature and Medicine During the Eighteenth Century*. London: Routledge.

Rosner, L. (1991) *Medical Education in the Age of Improvement. Edinburgh Students and Apprentices 1760–1826*. Edinburgh: Edinburgh University Press.

Rousseau, G.S. (1991*a*) *Enlightenment Borders: Pre- and Post-Modern Discourses: Medical, Scientific*. Manchester: Manchester University Press.

Rousseau, G.S. (1991*b*) Praxis 2: pineapples, pregnancy, pica and *Peregrine Pickle*. In Rousseau (1991*c*), 176–201.

Rousseau, G.S. (ed) (1991*c*) *The Languages of Psyche: Mind and Body in Enlightenment Thought*. Berkeley: University of California Press.

Rousseau, G.S. and Porter, R. (eds) (1980) *The Ferment of Knowledge*. Cambridge: Cambridge University Press.

Schofield, R.E. (1963) *The Lunar Society of Birmingham: A Social History of Provincial Science and Industry in Eighteenth-Century England*. Oxford: Oxford University Press.

Sekora, J. (1977) *Luxury: The Concept in Western Thought, Eden to Smollett*. Baltimore: The Johns Hopkins University Press.

Seymour, E.U.H. (1992) Bodying forth the mind: mind, body and metaphor, 1590–1640. Cambridge University, Ph.D. thesis.

Sheets-Pyenson, S. (1990) New directions for scientific biography: the case of Sir William Dawson. *History of Science*, **28**, 399–410.

Shortland, M. (1987) Screen memories: towards a history of psychiatry and psychoanalysis in the movies. *The British Journal for the History of Science*, **20**, 421–52.

Silvette, H. (1967) *The Doctor on Stage. Medicine and Medical Men in Seventeenth Century England* (ed. F. Butler). Knoxville: University of Tennessee Press.

Smith, W.D.A. (1982) *Under the Influence: A History of Nitrous Oxide and Oxygen Anaesthesia*. London: Macmillan.

Spacks, P.M. (1976) *Imagining a Self: Autobiography and Novel in 18th Century England*. Cambridge, MA: Harvard University Press.

Stannard, D.E. (1980) *Shrinking History: On Freud and the Failure of Psychohistory*. New York and Oxford: Oxford University Press.

Stansfield, D.A. (1984) *Thomas Beddoes M.D. 1760–1808, Chemist, Physician, Democrat.* Dordrecht: Reidel.

Stansfield, D.A. and Stansfield, R.G. (1986) Dr Thomas Beddoes and James Watt: preparatory work 1794–96 for the Bristol Pneumatic Institute. *Medical History,* **30,** 276–302.

Stengers, J. & Van Neck, A. (1984) *Histoire d'une Grande Peur: La Masturbation.* Brussels: University of Brussels Press.

Sterne, L. (1967) *The Life and Opinions of Tristram Shandy,* (ed. C. Ricks). Harmondsworth: Penguin.

Stock, J.E. (1811) *Memoirs of the Life of Thomas Beddoes MD.* London: J. Murray.

Sutherland, L.S. and Mitchell, L.G. (eds.) (1986) *The History of the University of Oxford,* vol. v, *The Eighteenth Century.* Oxford: Oxford University Press.

Thompson, E.P. (1963) *The Making of the English Working Class.* Harmondsworth: Penguin.

Timbs, J. (1876) *Doctors and Patients; or Anecdotes of the Medical World and Curiosities of Medicine.* London: Bently.

Todd, A.C. (1967) *Beyond the Blaze, A Biography of Davies Gilbert (Giddy).* Truro: D. Bradford Barton.

Trautmann, J. (1982) Can We Resurrect Apollo? *Literature and Medicine,* 1, 1–17.

Trautmann, J. & Pollard, C. (comp.) (1975) *Literature and Medicine: Topics, Titles and Notes.* Philadelphia: Society for Health and Human Values.

Trotter, T. (1989) *An Essay on Drunkenness.* London: Routledge [Reprint. 1st edition, 1804].

Trotter, T. (1807) *A View of the Nervous Temperament, A Practical Enquiry into the Increasing Prevalence, Prevention, and Treatment of those Diseases Commonly Called Nervous, Bilious, Stomach and Liver Complaints; Indigestion; Low Spirits; Gout etc.* London: Longman.

Webster, C. (1983) The historiography of medicine. In Corsi and Weindling (1983), 29–43.

Webster, C. (1986) The medical faculty and the physic garden. In Sutherland and Mitchell (1986), 683–724.

Weisz, G. (1988) The self-made mandarin: the éloges of the French Academy of Medicine, 1824–47. *History of Science,* 26, 13–40.

Wellman, K. (1992) *La Mettrie: Medicine, Philosophy, and Enlightenment.* Durham, NC: Duke University Press.

Welsford, E. (1935) *The Fool: His Social and Literary History.* London: Faber.

Wiener, P.P. (ed.) (1973) *Dictionary of the History of Ideas,* 4 vols. New York: C. Scribner's Sons.

White, W. (1939) The biographical essays of Sir William Osler and their relation to medical history. *Bulletin of the History of Medicine,* 7, 28–48

Williams, R. (1958) *Culture and Society: Coleridge to Orwell.* London: Chatto & Windus.

Wilson, P.K. (1992) Surgeon 'turned' physician: the career and writings of Daniel Turner. University of London, Ph. D. thesis.

Yolton, J. (1956) *John Locke and the Way of Ideas.* Oxford: Oxford University Press.

Young, R.M. (1973) Association of Ideas. In Wiener (1973), vol. 1, 111–18.

Young, R.M. (1988) Biography: the basic discipline for human science. *Free Associations*, 11, 108–30.

Young-Bruehl, E. (1994) A history of Freud biographies. In Micale and Porter (1994), 157–73.

9

The scientist as patron and patriotic symbol: the changing reputation of Sir Joseph Banks

JOHN GASCOIGNE

Science no less than religion needs its gallery of saints as sources of emulation to provide a sense of continuity and tradition. But, inevitably, posterity is selective in drawing up such a roll-call of the blessed as the past is scavenged for figures who seem best to conform to the needs of the present. Scientists of the nineteenth and twentieth centuries accorded most respect to the founding fathers of their discipline who left their mark in the manner most familiar to scientists of a later age. In seeking such a patrimony, science has understandably overlooked the claims of Joseph Banks, who published very little and left no indelible mark on any scientific discipline. But, as science has grown in size, complexity and expense so, too, the scientific estate has come increasingly to value the role of its patrons, protectors and paymasters. As a consequence, more attention has been accorded of late to statesmen of science like Francis Bacon. There are also a few signs that the significance of Banks as a promoter of science is beginning to be recognised, the most notable being the recent appearance of the first fully researched biography of Banks by H.C. Carter – a work which will serve as a precursor to an edition of his papers.

Perhaps, in time, posterity will belatedly accord to Banks the recognition that Cuvier bestowed in his éloge on behalf of the Academy of Sciences, that though 'the works which Banks left were few and their importance not much greater than their extent, nonetheless his name will shine with brilliance in the history of science'.[1] For Cuvier, schooled in the French tradition of the

[1] Cuvier (1827), 49.

scientist as a servant of the State, could recognise more than Banks's compatriots the importance of Banks as a protector and advocate for science both in his own country and in the relations between states.

The vicissitudes of Banks's reputation tell us something about the changing perceptions of the scope of science. But Banks's changing reputation is also of interest in another, albeit more localised, context. Of all Banks's multifarious activities that which – at least until recently – left the strongest mark on posterity was his role in the founding and maintenance of Australia as a European settlement. In a country in need of founding fathers Banks offered an attractive alternative to the convicts and their gaolers who were the first European settlers. Banks embodied many of the more palatable aspects of European settlement. His encounter with Australia in 1770 was associated with the cultivation of science both on account of the observation of the transit of Venus which had originally taken Cook's *Endeavour* to the South Seas and the pursuit of natural history commemorated in the name Botany Bay. He acted as the infant colony's main protector and advocate with the British Government and helped to promote its economy through the introduction of new plants and animals, of which the most notable was to be the Merino (though here Banks was slower to recognise its significance than John Macarthur for whom he had little respect). Banks, too, assisted the infant colony in promoting its external trade in a period when the East India Company looked with jealousy on New South Wales as a potential interloper in its trading empire. For all these reasons Banks has figured prominently in attempts to create a national Australian pantheon, his star ascending chiefly in periods when the need was strongest to define Australian identity.

However, Banks *qua* scientist and Banks *qua* icon of Australian nationhood have, until recently, remained in separate orbits, for the study of Banks in his Australian guise was largely a localised pursuit which had little impact on the wider world. Late nineteenth- and twentieth-century Australia might revere Banks but this did little to dispel the scientific oblivion to which he was consigned as one who was not a discoverer whose name could be attached to a scientific law or natural phenomenon.

When Banks died in 1820 the obituaries due to him as the Royal Society's longest-serving President and as a confidant of the great gave little hint of the oblivion which was to be his fate for most of the nineteenth century. The *Philosophical Magazine* concurred with Cuvier in according him high stature as a protector of science, averring that 'To the nation he has bequeathed . . . a name that it will never cease to cherish while science is encouraged or respected'.[2] Of Banks's tenure as President of the Royal Society *The Gentle-man's Magazine* wrote that 'never perhaps has it been filled with more honour to the individual, or more advantage to the interests of science'.[3] The *New Monthly Magazine* was even louder in its praises of Banks's services to the onward march of science claiming that 'Not even excepting the great Swedish Naturalist [Linnaeus], it may with justice be asserted, that Sir Joseph Banks was the most active philosopher of modern times'.[4]

But Banks's star was soon to fade in a century which, more and more, reserved its scientific plaudits for the discoverers rather than the patrons and facilitators of science. Moreover, as patron Banks was viewed as the embodiment of an old unreformed scientific order which depended on patronage and which was subordinated to Banks's personal whims. Hence the caustic comment of Humphry Davy, Banks's successor as President of the Royal Society, that 'he requested to be regarded as a patron and readily swallowed gross flattery . . . In his relations to the Royal Society he was too personal and made his circle too like a court'.[5] And, as the tide of reform engulfed the Royal Society and other scientific institutions along with the constituted order in Church and State,[6] so Banks came more and more to be regarded as part of a bygone and superseded old regime. Such an image of Banks as epitomising a form of scientific 'Old Corruption' was left largely unchallenged since the massive *Life and Letters*, which would normally have been due to a man of Banks's stature, failed to materialise. In the first place would-be biographers were deterred by the sheer mass of Banks's papers which included correspondence which may have run to something like 100 000 letters.[7] Secondly, from the 1880s the possibility of writing such an authoritative biography receded with the sale by Banks's descendant, Lord Brabourne, of many of his papers. Such manuscripts realised only derisory sums, an indication in itself of the obscurity into which Banks's reputation had sunk.[8]

For much of the nineteenth century, Banks remained little more than a name on the roll-call of past presidents of the Royal Society. Occasionally, one of his fellow naturalists would attempt to revive some interest in him as a major patron of science while apologetically acknowledging his deficiencies

[2] Anon. (1820*a*), 46.
[3] Anon. (1820*b*), 574.
[4] Anon. (1820*c*), 185.
[5] O'Brian (1987), 299.
[6] Foote (1951).
[7] Dawson (1958), xxviii.
[8] On the history of the dispersal of Banks's papers see Carter (1987), 15–23, Dawson (1958), xiii–xvii, Beaglehole (1962), 130–7 and Mander Jones (1949).

in the realm of scientific discovery. In 1835, for example, Sir William Jardine in his vast *Naturalist's Library* referred to him as someone who, though he did not belong to 'that exalted rank as a practical naturalist, by which Linnaeus and Cuvier have been distinguished' and who 'as an author may be said to be almost unknown', nonetheless, warranted a prominent place in the history of science since 'there have been few men in this country to whom physical science is more beholden, as his whole life was devoted to the encouragement, and his ample fortune to the illustration of it in all its branches'.[9]

Significantly, however, the most unequivocal nineteenth-century champion of Banks's place within the scientific hall of fame was a politician and promoter of scientific education rather than a practising scientist. For Banks's most eloquent nineteenth-century defender was the polymath, Lord Brougham, son of Banks's old Etonian school-friend, who included a long and laudatory account of Banks's achievements in his *Lives of Men of Letters and Science*, first published in 1845. With his customary skills as an advocate Brougham, like Cuvier before him, mounted the case for Banks's importance in the history of science despite the fact that he had contributed little of significance to any scientific discipline. 'It is rare to observe a name among the active and successful promoters of science', wrote Brougham, 'and which yet cannot easily find a place in its annals from the circumstance of its not being inscribed on any work or connected with any remarkable discovery'. But, as a politician himself, Brougham could recognise that science needed not only its experimenters and authors but also its patrons and protectors who could marshal the financial and political support to make scientific advance possible. And this, Brougham insisted, was Banks's great achievement as President of the Royal Society: 'no one, either before or since his time, ever occupied the high station in which he was placed with such eminent advantage to the interests of the scientific world'.[10] Brougham also mounted a spirited defence of Banks's conduct as President of the Royal Society in the disputes of 1783–4 which had given rise to the image of Banks as 'The Autocrat of the Philosophers'.

Brougham's forceful defence did little to sway the widely prevalent view of Banks as a scientific nonentity with autocratic tendencies. Such a view was both reflected and perpetuated in the nineteenth century's official register of

[9] Jardine (1835), 1, 17.
[10] Brougham (1846), 336–7.

the deeds of Britain's great, *The Dictionary of National Biography*. In his 1885 article on Banks, Jackson accorded some tepid praise to Banks as a scientific patron by writing that 'The character which Banks has left behind him is that of a munificent patron of science rather than an actual worker himself' but did little to develop the point adding simply that 'His own writings are comparatively trifling'.[11]

Some corrective to this diminished view of Banks's significance came with the publication of Banks's *Endeavour* journal. For it was a work which helped to re-establish Banks's credentials as a practising naturalist in an age when discovery still remained the chief gauge of scientific worth. It also served as a reminder of the way in which Banks had linked the cause of science with the expansion of British naval and imperial power – a topical issue in a Britain increasingly preoccupied with its imperial status. This edition was the work of J. D. Hooker, the son of Banks's young protégé, William Hooker, father and son both serving as Director of the Royal Botanic Gardens at Kew – an institution which had largely been Banks's creation.

By modern scholarly standards Hooker took a relaxed view of his duties as editor: apart from a brief introduction his contribution consisted of marking in red on the manuscript of Banks's journal at Kew which sections should be excised or rewritten. Nor did Hooker base his work on an original manuscript (though perhaps this was as well given his propensity to mark up the manuscript itself) for the source he used was a copy commissioned by Dawson Turner after Banks's death.[12] For all its faults, however, Hooker's edition did prompt some revival of interest in Banks and some reappraisal of the increasingly obscure role to which he had been consigned. Hooker himself seems to have been surprised how far the journal revealed Banks as a practising botanist of some accomplishment. As he wrote to the botanist, Baron Ferdinand Mueller, in distant Melbourne: 'The valuable feature of this work is the revealing of Banks in his right place as a working naturalist, the pioneer of the illustrious band of Natural Voyagers of which Darwin is the culminant. It nowhere appears in the accounts of Banks's life and works that he was a bona fide naturalist . . . Had he but published his collections what a mark he would have made in the scientific world proper'.[13]

Hooker once again emphasised in his preface that Banks merited inclusion

[11] Jackson (1885).
[12] Beaglehole (1962), 144–5.
[13] Royal Historical Society of Victoria. Box 6/4, Hooker to Mueller, 2 March 1896.

in the company of British scientists both for his own botanical work and because he served as an exemplar for future naturalist-explorers: 'My principal motive for editing the Journal . . . is to give prominence to his indefatigable labours as an accomplished observer and ardent collector, and thus to present him as the pioneer of those naturalist voyagers of later years, of whom Darwin is the great example'.[14] Hooker did have some success in persuading his contemporaries that Banks was entitled to greater and more favourable attention. Thus the *Spectator* in its review of the work wrote that 'the remarks of his able editor' had established 'That Sir Joseph's name and labour deserve recalling is abundantly evident'.[15]

The interest in Banks that had been reawakened by Hooker's edition of the *Endeavour* journal was strengthened by three publications which followed over the next decade and which further underlined Banks's significance both as a practising naturalist and as a promoter of imperial science: J. Britten's *Illustrations of the Botany of Captain Cook's Voyage Round the World in* HMS Endeavour *in 1768–71 by the Right Honourable Sir Joseph Banks, and Dr Solander* . . . Part One *Australian Plants* (1900–5) — a work which went some way towards putting in the public domain the botanical illustrations from the *Endeavour* expedition which Banks had left unpublished after lavishing so much expense and care on them — J.H. Maiden's *Sir Joseph Banks: The 'Father of Australia'* — of which more hereafter — and Edward Smith's *The Life of Sir Joseph Banks, President of the Royal Society* . . . (1911). The first two of these, together with Hooker's edition of the *Endeavour* journal, prompted the *Gardeners' Chronicle* of 1909 to remark of Banks that 'the results of his activity are not as widely known as they deserve to be; for he was something more than a patron of botany and horticulture, and an almost perpetual President of the Royal Society'. Indeed, the author continued, these recent works served 'to rehabilitate his fame as a great pioneer of science and colonisation'.[16] In short, it was again being appreciated that, despite the paucity of Banks's own published work, he was a major figure in the linking of British imperial expansion with the advancement of science.

Smith's biography which appeared in 1911 belatedly provided the basic record of Banks's life which, as a result of the chaotic disposal of his papers and the waning of public interest in him in the decades after he died, failed

---

[14] Hooker (1896), vii.
[15] Anon. (1897*a*), 374.
[16] Anon. (1909), 248.

to appear in the nineteenth century. Just how far public interest in Banks had declined was evident in the obstacles with which Smith had to contend in order to publish his work: for, after having been rejected by twelve publishers, it was finally accepted by Bodley Head only on condition that it be drastically abridged and that it include a section on caricatures of Banks which, it was hoped, might awaken greater interest (and sales) in such a little-known figure.[17] Ironically, this attempt to pitch the book at a popular level led one reviewer to dismiss it as 'a treasury of innocent gossip' and to urge the need for a more serious and comprehensive biography.[18] Smith also attempted to justify his attention to such a largely forgotten figure by linking his biography with a study of other features of the late eighteenth century, most notably the growing preoccupation with public duty. Thus Smith described Banks as 'a man of unbounded Public Spirit. Science was his passion, and the public service through the applications of science was his constant aim' even though 'The man has practically vanished from our ken, as an individual'.

Smith, then, provided further evidence for a reappraisal of Banks's significance in the history of science as someone who linked science to the political processes and public concerns of his day — hence his praise for Banks as an individual who 'gave enormous impulse to the study of Natural Science, and to the improvement of social conditions'.[19] But such a view of Banks as a statesman of science in the tradition of Bacon failed to make much ground despite Smith's advocacy. Raymond Pearl, an American reviewer of Smith's work, almost drew the opposite conclusion, implying that Banks's relative obscurity was a natural and not altogether unmerited consequence of his failure to publish any significant scientific work. 'The career of Banks', wrote Pearl, 'illustrates in a striking and complete way that fame which rests on anything other than solid achievement is a very fleeting sort of thing'.[20]

Though the works of Hooker, Britten, Maiden and Smith did do something to make the case for Banks as a major figure in both Britain's scientific and imperial history interest in him again began to wane after the first decade of the twentieth century. In part this may have been due to the fact that the general sketch of his significance that these works had provided could not be

---

[17] Lysaght (1964).
[18] Anon. (1911), 450.
[19] Smith (1911), vii–viii.
[20] Pearl (1911), 256.

fleshed out properly because of the chaotic dispersal of Banks's papers.[21] But this situation also reflects the state of the historiography of science and history more generally. History of science still largely remained the record of scientific discovery leaving little room for patrons and promoters of science such as Banks. On the other hand, historians of imperialism with their focus on political structures and economic movements paid little attention to the way in which the promotion of science could be linked with the expansion of empire.

Banks's lowly place in the scientific pantheon was not really questioned in the papers which appeared in the *Proceedings of the Linnean Society*[22] on the occasion of the centenary of his death or in the rather ambivalent tribute accorded to him by John Griffith Davies in *Nature* in 1943, the bicentenary of his birth. Davies who was even rather doubtful as to 'Whether Banks is to be accounted a man of science, as this term is now understood' dismissing him as a 'botanizer' rather than a 'botanist'. Davies also reinforced Banks's reputation as 'the autocrat of the philosophers' describing him as 'a masterful man' with a coterie of close followers. He did, however, concede that Banks's behaviour in continuing to promote scientific interchange between nations which were at war entitled him to the epithet of being 'a great European'[23] – high praise at a time when Europe was once again tearing itself apart. It was a description echoed by J.W. Hunkin, Bishop of Truro, who, in the only other bicentennial tribute to Banks, commended him for having 'fully recognized the international fellowship of Scientists'. Taking a rather wider view of Banks's significance than that of Davies, with his orientation to the practices of professionalised science, Hunkin described Banks as 'one of the most notable public figures of the eighteenth century'.[24]

Like Cuvier before them, then, Davies and Hunkin did – however fleetingly – recognise Banks's significance as a figure on the wider European stage

---

[21] Thus, although Hallder Hermansson's *Sir Joseph Banks and Iceland* (1928) gave yet another indication of Banks's wide and varied reach by drawing together the material available in the British Museum and at Kew on Banks's extensive involvement in the affairs of Iceland his account could only be a partial one: his preface records his regrets at not being able to find Banks's Icelandic journal sold at public auction in 1886 and, unbeknown to him, recently purchased by McGill University; still more buried from view was much of Banks's Icelandic correspondence which lay in packing cases in the Sutro Collection in San Francisco having narrowly escaped destruction in the San Francisco fire of 1906.

[22] Jackson (1920), Rendle (1920), Britten (1920), and Woodward (1920). Banks's centenary also prompted a brief entry in the American *Science* which consisted chiefly of a summary of the papers given at the Linnean Society. Anon. (1920).

[23] Davies (1943), 182–3.

[24] Hunkin (1943), 271–5.

of the Republic of Letters. This was an aspect of Banks's significance which had dwindled away over the course of the nineteenth century when Britain's scientific and intellectual life had become more divorced from that of the Continent than it had been in Banks's day. The events of the twentieth century have, however, once again drawn attention to the importance of international diplomacy in science as in other fields of human endeavour—a realisation which has led to some renewed interest in Banks and his role in attempting to balance the scientific interests of Britain with those of the Republic of Letters in the context of the wars generated by the American and French Revolutions. By publishing much of Banks's correspondence with members of the French Academy of Sciences and its successor, the National Institute, at a time when France and Britain were at war de Beer underlined the significance of Banks in evaluating the eighteenth century's professed belief that there was a 'republic of letters' which transcended national loyalties.[25] Dupree has drawn attention to the limitations of Banks's devotion to such a Republic of Letters when it clashed with national loyalties and, more recently, Daston has discussed the limitations of the notion of the Republic of Letters itself.[26]

There have also been some other indications that historians of science and historians more generally have begun to appreciate the way in which studies of Banks and his copious correspondence might enhance our understanding of the scientific and imperial concerns of late Georgian society. As the author of the standard text on the relations between science and government in pre-World War II United States, Dupree has also provided a stimulating outline of the ways in which Banks could be regarded as the principal architect of late eighteenth-century Britain's 'science policy'.[27] Such a 'science policy' was inseparably intertwined with Banks's role as an adviser to the British government on the ways in which the expansion and intensification of empire could serve British economic interests — a subject the breadth and significance of which has been illustrated by Mackay.[28]

Banks's role in the scientific and institutional life of his age has been the subject of a number of articles by Miller who stresses the need to relate Banks's policies and priorities to his role as a representative of a landed class which looked to science to provide both useful knowledge and rational

[25] de Beer (1960).
[26] Dupree (1964), Daston (1990, 1991).
[27] Dupree (1957, 1984).
[28] Mackay (1985).

amusement – a view of science which was to be challenged in Banks's later years as the hold of the landed classes on London learned society weakened.[29] Berman's study of the Royal Institution – though it deals only briefly with Banks himself – also provides similar insights into the way in which the workings of scientific institutions of the late eighteenth and early nineteenth centuries reflected the outlook of a landed oligarchy.[30] Such studies indicate the way in which Banks – who, for so long, has tended to be dismissed as a nonentity when measured against the canons of professionalised science – is beginning to be recognised as a figure of significance in the history of science.

As it is more and more realised that the scientific realm extends not only to the laboratory or the field but also to the institutions which provide it with its financial support so, too, Banks's star is beginning to rise. The changing reputation of Banks is, then, something of an index of the extent to which the scientific estate and its historians have been aware of their wider social and political setting. When Banks died his obituarists both in Britain and France were still acutely conscious of the importance of patronage in science and of Banks's role in promoting it. Over the course of the nineteenth and twentieth centuries Banks fitted less neatly into the image of science constructed by disciplinary specialists who measured scientific significance in terms of properly refereed and widely cited publications. Since World War II, however, as the scale and expense of science has made its dependence on patrons and particularly government ever more manifest, Banks has slowly begun to assume greater significance – a process that should gain greater momentum as his papers become more accessible in published form.

Banks as an Australian patriotic icon

So much, then, for Banks in his capacity as patron and promoter of science – a subject which historians have yet fully to develop to take account of the increasing recognition of the significance of political and institutional factors in the development of science as an activity to which society gives it support and its resources. I turn now to the subject of Banks *qua* founder of Australia in its European guise – an area which has been accorded comparatively greater recognition, even though this has directed attention to only one out of the many different fields on which Banks and his energies impinged.

---

[29] Miller (1981, 1983, 1986, 1989).
[30] Berman (1978).

As a distant and somewhat precarious offshoot of British civilisation the early colony of New South Wales remained very conscious of those who had played a role in persuading the British government to establish it. Banks was remembered not only for his role on the *Endeavour* expedition but also for his testimony before the House of Commons in 1779 which helped in the establishment of New South Wales as a penal colony as well as for his role in acting as an intermediary between the colony's early governors and the British government. When news of Banks's death belatedly reached the distant colony it was duly published in the *Sydney Gazette* of 3 February 1821 and in the following year the Philosophical Society of Australia erected on the South Head of Botany Bay 'an Inscription to commemorate the First Landing of Captain Cook and Sir Joseph Banks'.[31] The ceremony was graced by the presence of Governor Brisbane, President of the Philosophical Society, a keen astronomer who had been appointed as governor in 1820 with Banks's support. The inscription emphasised the connection between the British arrival in Botany Bay and the advancement of science, underlining Banks's role in the promotion of the study of Nature through the arts of patronage:

> Under the Auspices of British Science,
> These Shores were Discovered by
> James Cook, and Joseph Banks,
> The Columbus and Maecenas of their Time
> This Spot once saw them Ardent
> In the Pursuit of Knowledge.[32]

Such associations between Banks, Australia, and the self-evidently noble and uplifting pursuit of science were to be a continuing theme in the history of a nation ashamed of its origins as a penal settlement. Indeed, the erection of this inscription prompted one colonial poet to comment on the ironic contrast between the scientific origins of the name, Botany Bay, and the dismal associations it had acquired as a term for the penal settlement: thus he wrote of New South Wales that it was

> . . . big with virtues (though the flow'ry name
> Which Science left it, has become a scorn
> And hissing to the Nations), if our Great
> Be wise and good.[33]

---

[31] *Sydney Gazette*, 15 March 1822.
[32] *Ibid.*, 22 March 1822.
[33] *Ibid.*

This theme of the links between Banks, Australia and the civilising virtues of science was again stressed in one of the first biographical sketches of Banks, a work written by George Suttor and published in Sydney in 1855. Suttor had emigrated to Sydney as a free settler with the support of Banks who, as Suttor wrote, 'was considered the father and Founder of the Australian colonies'.[34] In Suttor's account, Banks and his involvement with Australia becomes part of the global reach of the Enlightenment – thus he praised Banks as one 'who turned the attention of the human mind to the discovery of the hitherto unknown portions of our globe, [and] opened also new paths to the arts and sciences . . . in an age when the spread of knowledge progressed over the civilised nations of Europe'.[35]

Australian interest in Banks was much promoted in the late nineteenth century by the purchase from Lord Brabourne in 1884 of Banks's papers relating to Australia by Sir Saul Samuel, the New South Wales Agent-General in London.[36] The papers came to Sydney at a time when the country was becoming increasingly preoccupied with its origins and identity as a nation. The centenary of European settlement was shortly to be held in 1888 and the movement for the Federation of the six colonies into a single nation (which came to fruition in 1901) was already under way. Banks was an obvious candidate for the role of a founding father in a country still anxious to hold tight to its connection with British civilisation.

Unfortunately, however, Banks was not only linked to the uplifting pursuit of science but also with the origins of Australia as a penal colony – an aspect of their past that Australians were then anxious to play down so far as possible. It was a side of Banks on which G.B. Barton commented in his official history of New South Wales which appeared in 1889, a year after the centenary celebrations. Barton was in no doubt that it was Banks who was the colony's true father: 'The idea of founding a colony at Botany Bay', he wrote, 'clearly originated with Banks; it was proposed by him in 1779; from him it passed to others, and was at last formulated in set terms for the

---

[34] Mackaness (1948), 16.

[35] Suttor (1855), [3].

[36] The arrangement between Samuel and Brabourne was, according to Samuel, 'that in the event of any more papers being found by him relating to the same subject [Australia] they were to be the property of the New South Wales Government'. When Samuel found that Brabourne had in fact disposed of some papers bearing on Australian themes privately be attempted in vain to initiate legal action (Mitchell Library, Sydney, CYA 906, Parkes Correspondence, Vol. 36, pp. 224–6, Samuel to Parkes, 16 July 1898). Brabourne's actions were, however, made public in the *Athenaeum* since a letter of Samuel to Hooker on the subject was published in that journal in 1897 (*Athenaeum*, 1897, pp. 547–8).

approval of the Government by men who quoted Banks as the great authority on the subject'. Not only did he largely initiate its foundation but his paternal role also extended to sustaining it in its early days: there is no exaggeration in saying that, during the active parts of his life, no measure of any importance was adopted without his opinion having been taken on the subject'. But such claims to fatherhood had to be balanced against the bastard origins of the colony. 'In reading the evidence given by Banks before the Committee', Barton continued, 'it is not pleasant to find him identifying himself so readily with the proposal to establish a penal settlement at Botany Bay'. In seeking to exonerate Banks to some extent from such associations Barton took refuge in historical relativism arguing that 'when a man of his character becomes conspicuous among the patrons of a system which is now universally detested, it should not be forgotten that his views on the subject were not so much his own, as those of the age in which he lived'.[37]

Such a defence appears to have done something to cleanse Banks of the taint of initiating the colony as a penal settlement particularly since those who looked to him as a founding father could dwell on the more edifying subject of his scientific associations. The subject of the link between Banks and the convict system was thereafter largely ignored. Banks's links with the convict system did not, then, greatly diminish his attractions as a founding father — a role in some demand as Federation and full nationhood approached. Moves for the erection of a statue to Banks surfaced in October 1895[38] though with no more practical effect than later such attempts. In 1898 E.E. Morris delivered a lecture on Banks before the delegates to the conference of the Library Association of Australia praising him as 'the Maecenas of science' and 'the first suggester of the foundation of the colony of New South Wales'. 'The colony once founded', Morris continued, 'Banks did not cease his efforts on its behalf. It is no exaggeration to say that for thirty years he was the unpaid "Agent-General" for New South Wales'. He concluded by urging Sydney to erect a statue to one 'who more than any other man deserves the proud title of "Father of Australia" '.[39]

Such adulation of Banks intensified in the years after Federation. On 21 January 1905, a few days before the commemoration of the beginnings of European settlement on Australia Day, Sydney's major newspaper, *The Sydney Morning Herald*, carried an article entitled 'Australia's First, Best, and

[37] Barton, 1889, l–li, lvi.
[38] *Sydney Morning Herald*, 19, 25 October 1895.
[39] Morris (1898), 59.

Greatest Friend' which urged the erection of 'some worthy memorial' to Banks to whom 'Australia in general, and New South Wales in particular, owes . . . an undying debt of gratitude and affection'.[40] It was a view echoed a few months later in a speech by Dr Dixson, President of the Linnean Society of New South Wales, who maintained – using language that underlined Australia's fidelity to the British connection – that Australia owed Banks a statue since, though 'Cook's discoveries made Greater Britain possible', it was Banks who 'was the antidote to the unwisdom of the statesmen of the period'.

A few days later the cry was taken up by Banks's most influential Australian champion, J.H. Maiden, who, in a speech to the Australian Historical Society, argued that it was odd that a statue had not been erected to Banks when 'there were statues to men whose services to Australia were vastly inferior to those of Banks'.[41] Maiden's interest in Banks arose naturally out of his own professional concerns. Since 1896 he had combined the post of Director of the Sydney Botanical Gardens and Government Botanist and he was, moreover, an authority on the subject closest to Banks's heart, economic botany, including among his publications works such as *The Useful Native Plants of Australia*. Maiden was also an advocate of the need for public scientific education having served from 1894 to 1896 as New South Wales's Superintendent of Technical Education as well as having been active in promoting the Technological Museum and the Botanical Gardens as centres for public instruction.[42] His advocacy of Banks's claims as a national icon can be thus regarded as part of his work in promoting the scientific as well as the civic education of his compatriots.

With Maiden in the lead the campaign to institute some public monument to Banks gathered momentum. The following month 'An enthusiastic meeting was held . . . for the purpose of divising means for the erection of a memorial to perpetuate the memory of and services of Sir Joseph Banks'. It was a meeting which was chaired by the Governor and included the Premier, Mr J.H. Carruthers, and Sir Edmund Barton, Australia's first Prime Minister (from 1901 to 1903),[43] an august gathering which underlined the high place Banks enjoyed in the national pantheon. Out of the meeting emerged a committee with the duty of raising funds by public subscription with Sir Francis Suttor (grandson of Banks's protégé and biographer, George Suttor) serving as President and Maiden as Secretary.

---

[40] *Sydney Morning Herald*, 21 January 1905.
[41] *Ibid.*, 20 March 1905.
[42] Lyons and Pettigrew (1986), 381–3.
[43] *Sydney Morning Herald*, 26 May 1905.

Maiden's own efforts to advance the Banksian cause culminated in his publication of a work predictably entitled *Sir Joseph Banks: The 'Father of Australia'* published gratis by the New South Wales Government Printer in 1909 since all royalties from its sale were to go to the Banks Memorial Fund. It was an unusual publication consisting of a compilation of material relating to Banks rather than a fully digested biography. It was, of course, Maiden's central contention that Banks's 'services have not been adequately recognised either by Britain or by Australia. This neglect is a reflection on Australians who, however, have the excuse that they err in very good company'.[44]

After this flurry of activity the promotion of Banks as an Australian national icon lost much of its momentum. From 1914 all else was overshadowed by the war. Moreover, the blood sacrifice of Australian troops at Gallipoli provided an alternative focus for national sentiment to that associated with the nation's origins as a British colony. The Banks Memorial Fund remained in being but had become largely dormant well before Maiden's death in 1925. Occasional articles on Banks continued to appear such as one from 1919 entitled 'Sir Joseph Banks, the father of Australia: a great botanist who was one of the most public-spirited of men'[45] but the statue of Banks which Maiden had hoped to place in the vestibule of the Mitchell Library — the specialist Australian and Pacific wing of the Public Library of New South Wales — failed to materialise.[46]

In 1929 the Mitchell Library did, however, add to its holdings of Banksian manuscripts by purchasing at Sotheby's part of 'A very remarkable collection of letters addressed to Sir Joseph Banks 1743–1820 by many of the illustrious men of his day'. The fact that the Library had to pay the sum of £8600 for this purchase was an indication of the extent to which interest in Banks had increased since the Sotheby sale of Banks material in 1886 which had realised only a derisory amount.[47] The other part of this collection went to what became the Australian National Library in Canberra,[48] the federal capital, an indication of the extent to which Banks was identified with the foundation of Australia rather than simply New South Wales. The increasing bulk of Banksian material in the Mitchell Library prompted the publication in 1936 of a work by the Australian historian, George Mackaness, the purpose of which is captured by its title: *Sir Joseph Banks. His Relations with Australia.*

[44] Maiden (1909), xi.
[45] Shepherd (1919), 29–30.
[46] Mitchell Library, Doc. 2011, Maiden to Charles Hedley, 27 March 1909.
[47] Carter (1987), 23.
[48] Mackaness (1936), vii.

It was a work which went even further than Maiden's in highlighting the extent of Banks's involvement in overseeing the early colony of New South Wales. It also added further lustre to Banks's claims as a national secular patron saint: Mackaness wrote, for example, that 'Banks was the most disinterested of men. No suspicion of commercialism or self-seeking marks his association with New South Wales'.[49]

The approach of the bicentenary of Banks's birth in 1943 finally meant that the Banks Memorial Fund and J.H. Maiden's exertions on its behalf began to bear fruit. Though in 1943 Australia was distracted by other events there was still sufficient interest in Banks and an Australian memorial to him to prompt the passage of an act of the New South Wales parliament to regularise the situation of the moribund fund by establishing the Sir Joseph Banks Memorial Fund Trust. But the issue of how best to spend the money already accrued and the money that it was hoped would be forthcoming from government and the general public aroused controversy now that the original idea of erecting a statue was deemed 'out of keeping with modern thought'.[50] Interested parties canvassed on behalf of various other possible forms of memorial which included a museum in the National Herbarium, a lectureship in Australian history or a memorial at Cook's Landing Place at Botany Bay. But the most insistent and ultimately successful voice was that of the Royal Australian Historical Society which urged that the money be used for publication of Banks's papers. In support of such a project it was argued that Banks 'was the patron of scientific endeavour and exploration and displayed a keen interest in the colony's welfare. Yet no really satisfactory biography has been produced'.[51]

To meet the legal niceties of the situation another act was passed in 1945 terminating the Sir Joseph Banks Memorial Trust and transferring its funds to the Trustees of the Public Library of New South Wales – the body charged with preparing the editions of Banks's papers. In preparing this bill in the Parliament of New South Wales Mr Robert Heffron, then Minister for Education and a future Premier, took up the familiar theme of the association between Banks's enlightened activities and Australian nationhood: 'As President of the Royal Society, he was able to initiate and develop many of the early schemes that gave rise to our progressive Commonwealth, as it is

---

[49] *Ibid.*, 23.
[50] Mitchell Library, MSS 163, Box 1(1), Sir Joseph Banks Memorial Trust Minute Book, 19 August 1943.
[51] *Ibid.*, 28 October 1943.

to-day . . . So great was his influence, interest, and practical support at the time, that he has been aptly termed "The Father of Australia" '. The publication of Banks's work, it was hoped, would help promote Australia internationally by 'bringing our great Commonwealth before the notice of other nations'.[52] And indeed the Banks Memorial Trust made possible the publication of Beaglehole's definitive edition of Banks's *Endeavour* journal in 1962 and, in 1979, of Carter's edition of Banks's sheep and wool correspondence — a work which made apparent the importance of such little known repositories of Banks's papers as Yale University and the Sutro Library, San Francisco. After a long lull in the activities of the Trust its remaining funds are now being devoted to producing a digitally enhanced CD-ROM copy of Banks's correspondence in the Mitchell Library, a project which will bring the Trust to an end.

How far can the Trust's waning energies be regarded as an index of the declining fortunes of Banks as a national icon? Given that Banks is so closely associated with the British origins of European settlement in Australia one would expect that as the ties between Australia and Britain have loosened so Banks's place in the national pantheon would diminish accordingly — and there are some signs of such a correlation. In the period up to about the 1970s Banks's star still shined bright: as we have seen the Banks Memorial Fund sponsored a number of major projects and articles on Banks such as that by M.H. Ellis in *The Bulletin* of 1953 with the familiar title 'The Father of Australia' continued to appear.[53]

A rather more equivocal attitude to Banks was evident in an article which appeared in 1946 by John O'Donnell in the left-wing journal, *Progress*, at a time when the country was pre-occupied by the task of post-war reconstruction. O'Donnell used the occasion of Australia Day 1946 to argue that 'The myth of Banks as a patron-saint, a benign prophet needs to be debunked' maintaining that 'On the one hand he was active, progressive and scientific: on the other, indifferent, reactionary and bigotted'. Nonetheless, O'Donnell still regarded Banks — for all his associations with a landowning oligarchy — as providing a source of valuable inspiration to Australia in promoting 'the planned science of reconstructed Australia'.[54]

Interest in Banks was, no doubt, encouraged by the anglophilia associated

[52] Mitchell Library, Box 1(1), Second Reading Speech, Sir Joseph Banks Memorial Bill, Legislative Assembly, 27 September 1945.
[53] Ellis (1953).
[54] O'Donnell (1946), 25, 27.

with Sir Robert Menzies (Prime Minister, 1949–1966), the dominant political figure of the period and leader of the politically conservative Liberal Party.
Appropriately, however, Banks, who, in his own lifetime, strove to avoid
being identified with any particular political faction, also transcended the
political divide as a national symbol. It was the Labor administration in New
South Wales which established the Sir Joseph Banks Memorial Trust in 1945
and provided it with much of its funding and Evatt, the Labor Party's most
important intellectual, presided over the production of the Beaglehole edition of Banks's *Endeavour* journal as Chairman of the Board of Trustees of
the State Library of New South Wales. Indeed, Evatt launched Beaglehole's
edition with a speech in which he maintained that 'it is perhaps not too much
to say that without Sir Joseph Banks there would not have either been a State
of New South Wales or an Australian Nation as we know them'.[55]

After the publication of the Memorial Fund's second major project –
Carter's edition of Banks's sheep and wool correspondence – in 1979, however, interest in Banks as a founding father appears to have waned as Australia's economic and cultural ties with Britain also declined in the wake of
Britain's entry to the European Economic Community in 1973. One of the
few attempts to revive Banks's credentials as 'The Father of Australia' was
that by the prominent columnist and keen promoter of an Australian cultural
identity, Max Harris, in 1984. In an article entitled 'Who founded Australia?'
in the national newspaper, *The Australian*, Harris yet again made the case for
Banks as 'the founding father of the European-era Australia' on the grounds
that 'every nation needs a founding father. One does not want to be a bit of
a bastard among the community of nation-states'. In attempting to make
this case Harris focused on Banks's scientific credentials rather than his links
with Britain. Banks he portrayed as 'Australia's Ben Franklin' because of his
ability to combine science and public life – though Harris light-heartedly
suggested that the fact that Banks 'veritably embodied Britain's new-found
spirit of intellectual inquiry' would be 'a plus Australians will regard as a
minus'. This he attempted to balance by arguing that there was nothing
effete about Banks and that 'It could be said that Ned Kelly was as game as
Joe Banks'. In a similar tongue-in-cheek spirit Harris also attempted to
mount a case for Banks as an exemplar of contemporary concerns – a proto-
feminist because of his alleged role in regularising the way in which female

---

[55] Evatt (1962), 2.

convicts were allocated and a proponent of the multi-cultural society because of his proposals for interracial marriage in New South Wales.[56]

But this rather implausible advocacy appears to have availed little – the bicentenary of Australia's European foundation in 1988 passed with few signs of a renewed interest in Banks despite the appearance of Carter's authoritative biography. The recent decision to remove his portrait from Australia's five-dollar note attracted little or no protest in contrast to the outcry prompted by the replacement of Caroline Chisholm, a figure whose work among the early female emigrants tied her much more directly to national sentiment. The two hundred and fiftieth anniversary of Banks's birth on 13 February 1993 also virtually passed unnoticed. Whether the eventual appearance of a multi-volumed edition of Banks's correspondence and the scholarly industry that is likely to accompany it will generate a renewed attempt to link his achievements with the cause of Australian nationalism remains to be seen but it would appear doubtful.

As Australia moves further down the path to a republic the demand is for home-grown heroes and Banks's British associations are likely to lessen his appeal as a national icon. Moreover, Banks's scientific standing and his association with the culture of the Enlightenment which for so long provided a source of attraction in a nation shameful of its convict origins are less likely to be of such self-evident appeal in a world where science is no longer considered an unequivocal source of progress and human betterment.

Conclusion

The vicissitudes of Banks as an Australian national icon – as well as his fluctuating fortunes in the historiography of science and of the eighteenth century – suggest some more general remarks about the role of science and scientists in the culture of the age. As a promoter and patron of science rather than a practising scientist with a string of original publications to his name Banks did not conform to the public image of science and so has largely been written out of the scientific history of the period. Moreover, since he did not hold formal government office, his role in the more general history of the period has also been largely overlooked. The fact that scientists like other mortals need finance, social standing and political influence has only slowly

---

[56] Harris (1984).

and belatedly coloured the writing of the history of science so that Banks still remains something of a scientific Cheshire cat, largely invisible except for the influence he exerted – a situation which may change as more and more of the vast Banks archive is made accessible. In particular, Banks's importance as a promoter of schemes to link the advancement of science with the expansion of empire is receiving particular attention after being neglected by a historiography of science which, in Enlightenment fashion, tended to portray the quest for scientific knowledge as a corrective to such baser human instincts as the quest for domination.

But though the scientific profession and historians of science have tended to relegate Banks to the margins of their subject his identification with the promotion of science does much to explain his appeal as a focus of Australian nationalism in the late nineteenth and twentieth centuries when he was largely neglected within Britain and elsewhere. For in Australia the imperial dimensions of Banks's science served to connect a distant and insecure nation with both the British Empire and the commonwealth of learning. These connections between Banks, the foundation of Australia and scientific exploration such as the *Endeavour*'s role in observing the Transit of Venus and the activities of Banks and his clients in making available to an admiring scientific audience the novel flora and fauna of Australia are naturally something that Australians have tended to highlight as an antidote to the brute fact that Australia began as a penal colony. Banks's role in establishing the penal colony has tended to be glossed over while his role in integrating the early Australian colony into a world-wide network of collectors and in generally putting it on the scientific map has been highlighted.

The fact that Banks promoted such scientific activities as well as exercising unofficial supervision of the colony's early administration, without holding any formal government office, has perhaps added to his lustre as an Australian rather than a British hero, since it has served to divorce him somewhat from too close an identification with the British government. In his Australian capacity, then, Banks has been generally cast in the role of the scientist as hero — a role played in the United States, another nation with deep eighteenth-century roots, by Franklin and, to a lesser extent, Jefferson. In Australia — in contrast to the United States — the imprint of the eighteenth century and its Enlightenment culture was so overshadowed by memories of the squalor of the penal system that Banks's place in the national pantheon has never been secure. It is likely to become less so as Australians' conception of their nation as an outcome of British initiatives continues to wane. Ironically, at a time when historians of science have blazed more and more paths through the institutional and social history of science which are likely to lead to a greater recognition of Banks's significance, his role as a national icon — the role which has hitherto done most to save him from obscurity — appears likely to fade.

Bibliography

Anon. (1820*a*) Biographical memoirs of the late Right Honourable Sir Joseph Banks. *Philosophical Magazine*, 56, 40–6.

Anon. (1820*b*) Obituary. Sir Joseph Banks. *Gentleman's Magazine*, 90(1), 574, 637–8.

Anon. (1820*c*) Memoir of Sir Joseph Banks. *New Monthly Magazine*, 14(2), 185–94.

Anon (1897*a*) The journal of Sir Joseph Banks. *Spectator*, 78, 374–5.

Anon. (1909) Sir Joseph Banks. *Gardeners' Chronicle*, 46, 248–9.

Anon (1911) Review of W. Smith, *Life of Sir Joseph Banks*. *Nation*, 93, 450.

Anon. (1920) Centenary of Sir Joseph Banks. *Science*, n.s. 52, 123.

Barton, G.B. (1889) *The History of New South Wales from the Records, Vol.1, Governor Phillip 1783–89*. Sydney.

Beaglehole, J.C. (ed.) (1962) *The Endeavour Journal of Joseph Banks 1768–1771*, 2 vols. Sydney: Public Library of New South Wales in association with Angus and Robertson.

Berman, M. (1978) *Social Change and Scientific Organization. The Royal Institution 1799–1844*. London: Heinemann.

Britten, J. (1920) Banks as a botanist. *Proceedings of the Linnean Society*, Supplement, 132, 15–20.

Brougham, H.P.B. (1846) Sir Joseph Banks. In his *Lives of Men of Letters and Science who Flourished in the Time of George III*, 2 vols. London: C. Knight, 336–90.

[Brown, E.] (1905) Sir Joseph Banks. In his *What Australia Lacks*. Sydney: Eagle Printing House, 23–30.

Carter, H.B. (1979) *The Sheep and Wool Correspondence of Sir Joseph Banks 1781–1820*. Sydney: The Library Council of New South Wales; London: British Museum (Natural History).

Carter, H.B. (1987) *Sir Joseph Banks (1743–1820). A Guide to Biographical and Bibliographical Sources*. London: St Paul's Bibliographies in association with the British Museum (Natural History).

Carter, H.B. (1988) *Sir Joseph Banks 1743–1820*. London: The British Museum (Natural History).

Cuvier, G.L. (1827) *Recueil des Eloges Historiques*, 3 vols. Paris.

Daston, L. (1990) Nationalism and scientific neutrality under Napoleon. In T. Frängsmyr (ed.), *Solomon's House Revisited. The Organisation and Institutionalisation of Science*, Canton, MA: Science History Publications, 95–119.

Daston, L. (1991) The ideal and reality of the Republic of Letters in the Enlightenment. *Science in Context*, 4, 367–86.

Davies, J.G. (1943) Sir Joseph Banks, P.C., K.C.B., F.R.S. (1743–1820). *Nature*, 151, 181–3.

Dawson, W.R. (1958) *The Banks Letters. A calendar of the manuscript correspondence of*

Joseph Banks preserved in the British Museum, The British Museum (Natural History) and other collections in Great Britain. London: Trustees of the British Museum.

De Beer, G. (1960) *The Sciences were Never at War.* Edinburgh: Nelson.

Dupree, A.H. (1957) *Science and the Federal Government. A History of Policies and Activities to 1940.* Cambridge, MA: Harvard University Press.

Dupree, A.H. (1964) Nationalism and science – Sir Joseph Banks and the wars with France. In D.H. Pinkney and T. Ropp (eds), *A Festschrift for Frederick B. Artz*, Durham, NC: Duke University Press, 37–51.

Dupree, A.H. (1984) *Sir Joseph Banks and the Origins of Science Policy.* Minneapolis: The James Ford Bell Lectures, no. 22, University of Minnesota.

Ellis, M.H. (1953) The father of Australia. *The Bulletin*, **74**, 25.

Evatt, H.V. (1962) *Notes for an Address by the President of the Trustees of the Public Library of New South Wales, 15 February 1962.*

Foote, G.A. (1951) The place of science in the British Reform Movement 1830–50. *Isis*, **42**, 192–208.

Harris, M. (1984) Who founded Australia?. *The Australian*, 28–9 July, Weekend Magazine, 4.

Hermannsson, H. (1928) *Sir Joseph Banks and Iceland.* In *Islandica. An Annual Relating to Iceland and the Fiske Icelandic Collection in Cornell University*, **18**.

Hooker, J.D. (ed.) (1896) *Journal of the Right Honourable Sir Joseph Banks during Captain Cook's First Voyage in H.M.S. Endeavour.* London.

Hunkin, J.W. (1943) Bicentenary of Sir Joseph Banks. *The Fortnightly*, **160**, 271–5.

Jackson, B.D. (1885) Sir Joseph Banks. In S. Lee and L. Stephen (eds), *Dictionary of National Biography*, vol. 3, 129–33.

Jackson, B.D. (1920) Sir Joseph Banks as a traveller. *Linnean Society*, **132**, Supplement, 9–15.

Jardine, W. (1835) Memoir and portrait of Joseph Banks. In his *Naturalist's Library, Icthyology*, vol. 1. Edinburgh, 17–28.

Lyons, M. and Pettigrew, C.J. (1986) J.H. Maiden. In B. Nairn and G. Searle (eds.), *The Australian Dictionary of Biography*, Melbourne, vol. 10, 381–3.

Lysaght, A.M. (1964) A grangerised copy of Edward Smith's *Life of Sir Joseph Banks. Journal of the Society for the Bibliography of Natural History*, **4**, 206–9.

Mackaness, G. (1936) *Sir Joseph Banks: His Relations with Australia.* Sydney: Angus and Robertson.

Mackaness, G. (1948) *Memoirs of George Suttor, F.L.S. Banksian Collector (1774–1859).* Sydney: privately published.

Mackay, D. (1985) *In the Wake of Cook. Exploration, Science and Empire, 1780–1801.* London: Croom Helm.

Maiden, J.H. (1909) *Sir Joseph Banks: The 'Father of Australia'.* Sydney: W.A. Gullick; London: Kegan Paul.

Mander Jones, P. (1949) *History of the Papers of Sir Joseph Banks.* Mitchell Library, MS Ab 67–9/7.

Miller, D.P. (1981) Sir Joseph Banks: An historiographical perspective. *History of Science*, **19**, 284–92.

Miller, D.P. (1983) Between hostile camps: Sir Humphry Davy's Presidency of the Royal Society of London, 1820–1827. *British Journal for the History of Science*, **16**, 1–47.

Miller, D.P. (1986) Method and the 'micropolitics' of science: the early years of the Geological and Astronomical Societies of London. In J.A. Schuster and R. Yeo (eds), *The Politics and Rhetoric of Scientific Method. Historical Studies*, Dordrecht: Reidel, 227–58.

Miller, D.P. (1989) 'Into the valley of darkness': reflections on the Royal Society in the eighteenth century. *History of Science*, 27, 155–66.

Morris, E.E. (1898) Sir Joseph Banks. *Transactions and Proceedings of the Library Association of Australia*, October, 51–9.

O'Brian, P. (1987) *Joseph Banks. A Life*. London: Collins Harvill.

O'Donnell, J. (1946) Sir Joseph Banks and Australian science. *Progress*, 1, 24–7.

Pearl, R. (1911) An eighteenth-century patron of science. *Dial*, 51, 255–6.

Rendle, A.B. (1920) Banks as a patron of science. *Proceedings of the Linnean Society of London*, 132, Supplement, 9–15.

Shepherd, S. (1919) Sir Joseph Banks, the father of Australia: a great botanist who was one of the most public-spirited of men. *The Lone Hand*, 29–30.

Smith, E. (1911) *The Life of Sir Joseph Banks, President of the Royal Society, With Some Notices of His Friends and Contemporaries*. London: John Lane, The Bodley Head.

Suttor, G. (1855) *Memoirs Historical and Scientific of the Right Honourable Sir Joseph Banks*. Parramatta.

Weld, C.R. (1848) *A History of the Royal Society, with Memoirs of the Presidents*, 2 vols. London.

[Woodward, A.S.] (1920) Banks as a trustee of the British Museum. *Proceedings of the Linnean Society*, 132, Supplement, 20–1.

10

Metabiographical reflections on Charles Darwin

JAMES MOORE

> If we suppose that what is produced in cultural practice is a series of
> objects, we shall . . . set about discovering their components. . . . But I
> am saying that we should look not for the components of a product but
> for the conditions of a practice. When we find ourselves looking at a
> particular work, or group of works, often realising, as we do so, their
> essential community as well as their irreducible individuality, we should
> find ourselves attending first to the reality of their practice and the
> conditions of the practice as it was then executed.
>
> Raymond Williams[1]

Navel-gazing is not a noted source of historical insight but a personal perspective is essential to this chapter: in 1991 Adrian Desmond and I published a
best-selling biography of Charles Darwin. In reflecting on our collaboration
I intend that a biographical, not an egotistical, spirit should prevail.[2]

An early reviewer of our *Darwin* felt moved to associate it with a text from
that famous study of modern industrialism, Roald Dahl's *Charlie and the
Chocolate Factory*.[3] Here is the scene witnessed by the book's hero Charlie
Bucket in the proprietor's 'Inventing Room':

> Mr Wonka himself had suddenly become even more excited than
> usual, and anyone could see that this was the room he loved best of
> all. He was hopping about among the saucepans and the machines

---

[1] Williams (1973), 16.

[2] Many thanks to Adrian for jogging my memory and reading this chapter in draft. The interpretations are my own.

[3] Neve (1991).

like a child among his Christmas presents, not knowing which thing to look at first. He lifted the lid from a huge pot and took a sniff; then he rushed over and dipped a finger into a barrel of sticky yellow stuff and had a taste; then he skipped across to one of the machines and turned half a dozen knobs this way and that; then he peered anxiously through the glass door of a gigantic oven, rubbing his hands and cackling with delight at what he saw inside.

What stirred our reviewer's childhood memories of this passage? On page 649 of *Darwin* he found our own Charlie in his study, littered with glass-covered pots full of worms. He stumbles around at night, flashing lights at them. He lifts the lids to check their senses of temperature, touch and taste. Whistles are blown, bassoons tooted, and the piano played to test their hearing, and Darwin peers anxiously through the glass to see how clever they are at burying scraps of paper. 'Rubbing his hands and cackling with delight' — as it were — he discovers that intelligence *is* there.

Getting Charlie Darwin and Charlie Bucket bracketed together in the *Times Literary Supplement* is no mean achievement, even if I say so myself. Adrian and I must have done something right, and I want to reflect first on what that might be.

Darwin among the biographers

My emeritus colleague Nick Furbank says that biographies are published nowadays like novels were when he was young. The *Irish Times* obliged in 1991 by announcing the autumn lists under the headline 'Novelists out-gunned by the biographers'. Nick, with the best life of E. M. Forster to his credit, explains the turnabout by consumer preference. Fiction can be frightening, real lives are safe. The life of X is an easier read than X's works (the '*Reader's Digest* syndrome'). And only real-life hijinks satisfy the voyeur. 'Unauthorised' Madonnas and lurid Philip Larkins are always being launched. Professional writers and their agents make a living from it.

Not professors, *pace* Nick. They write for tenure and promotion. Since the 1950s academic biographers — Americans too often — have churned out huge, fancy-footnoted tomes, full of mind-numbing detail. In these the subject is treated as an indexed card on which the professor has noted down certain facts that, on grounds of proven scholarship, it is necessary to trot out. Since

1978 there has even been an 'interdisciplinary quarterly', *Biography*, to facilitate such work.

Against this upsurge of dusty display, fatal to subject and reader alike, a seasoned biographer protested in *The Times*:

> Research does not teach you about people, and it is people who
> bring biography alive. Biographers have become like morticians
> awaiting the arrival of the next body on the slab. Each person is
> given the same standard treatment – perfectly acceptable in terms
> of research, but lifeless in narration – and 'biography' conversation
> has become a *danse macabre* among the 'stiffs'. 'Oh, he's doing
> Cecil Beaton', or 'Oh, she's on Nancy Mitford – a biggie'. 'But
> hasn't she been done?' or 'It's high time someone got at Gerald
> Berners, but I gather they won't release the papers'.

Biography should be an 'exciting read,' the disgruntled critic ended. 'Must the academics really have it all their own way, and does the art form no longer stand a chance?'[4]

Regaled and admonished, I took this lament to heart long before *Darwin* was begun. How relevant it still is. In recent years literary biographies by professionals scooped the best awards: Holroyd's Shaw, Holmes's Coleridge, Richardson's Picasso, Ackroyd's Dickens. Lives by academics suffered by comparison, even Shelden's Orwell and Kemp's Ruskin. And little wonder. Consider Ian Ker's Newman: it begins, 'John Henry Newman was born on . . .' and proceeds to unfold events almost as if they had occurred entirely within men's minds. Or Rosemary Ashton's George Henry Lewes: full of scholarly exegesis, it is too cautious to make the imaginative leap to a engaging narrative, as literary biographies do. Or John Worthen on D. H. Lawrence's early life: lucid, judicious, encyclopaedic, his volume is the first of three, each by a different scholar. Together they are expected to achieve a greater realism and 'give the lie . . . to the idea that any single view, however detailed and comprehensive, could ever be "definitive" '.

Parenthetically, I agree with the *TLS* reviewer that this justification for multiple authorship is a mistake. The main point of writing biography is of course 'not to be definitive, but to try to understand; and to claim that different pairs of eyes convey a more realistic picture is to misunderstand the

---

[4] Blow (1987).

endeavour. The skill is in the biographer's capacity for split vision', an ability to judge her subject with one eye and see as her subject sees with the other. 'The best biographical writing is the distillation of two minds, one of which sympathizes with, judges and displays the other.'[5]

Or perhaps the two minds may belong to joint-authors who view their subject in different but complementary ways. The example of *Darwin* would suggest so, though such a life is rare.

Most scientific biographies of late have been written both by and for scholars. Some are magisterial tomes evincing vast research, such as Westfall's Newton and Cardwell's Joule; others are modest studies, with only scant claim to rank as lives. Robert Stafford's Murchison has a single biographical chapter followed by detailed accounts of the 'King of Siluria's' geological conquests, region by region. Geoffrey Cantor's fine Faraday purports to be a biography, though its chapters are mainly analytical essays on Faraday's physics and religion. And what about Crosbie Smith and Norton Wise's 'biographical study' of Lord Kelvin, the most impressive collaborative scientific life to date? Brilliantly, it traverses Thomson's career again and again, treating every aspect of his technical and scientific work in 650 pages. This is the book's rich meat; the wrappers consist of 160 pages of more-or-less straightforward biography. All in all it is a sumptuous sandwich – for scholars. The book will never appeal beyond a small circle of well-heeled specialists and university libraries.

Indeed, all these biographies come replete with one or more of the following: a forbidding price-tag, intimidating footnotes, rambling primary-source quotations, distracting historiographic asides, jarring first-person interventions ('in this chapter I shall argue . . .'), and a confusing narrative structure. The academic appeal and permanent value of these works is beyond doubt, and the Smith-Wise volume is as outstanding in its way as Westfall's Newton. I only wish to point out that one must be highly motivated to tackle them. And one is more likely to come away staggered by their scholarship than moved or entertained.

The situation with recent biographies of Charles Darwin is curiously reversed. Here popular or literary works hold the field. Irving Stone's *The Origin* (1980) is suggestive and insightful for a 'biographical novel', Peter Brent's *Charles Darwin* (1981) charming but sketchy, and Ronald Clark's *The Survival of Charles Darwin* (1984) only half-biography (and not the author's

---

[5] Judd (1991).

best). John Bowlby's *Charles Darwin* (1990) is good on psychology, poor on other sciences, and an indifferent read. It is however the best life of the lot, drawing on the latest research.[6]

Why have academic biographers neglected 'the greatest naturalist since Newton'? I suspect the reason has to do with the scary scope of Darwin studies. Since the mid- 1980s scholars have fled in awe from the avalanche of notebooks, texts, and marginalia poured out by the 'Darwin industry'. Peter Bowler, a veteran historian, is no exception. His *Charles Darwin* (1990), the first in a series of 'Blackwell Science Biographies', disclaims in its opening paragraph being 'a biography of Darwin in the conventional sense'. The book is in fact a sketch of Darwin's life to 1859, followed by a few chapters outlining post-Darwinian debates; a fine synopsis and synthesis of recent scholarship, especially useful for students. 'A really detailed biography of the kind that is now becoming possible', Bowler warns, 'must necessarily be an enormous volume beyond the reach of many non-specialists'.[7]

Just the sort of book that made the *Times*'s critic wonder whether biography as an 'art form' stands a chance. But why should a profusion of primary sources rule out a literary biography of Darwin when even larger archives (like Bernard Shaw's) have sustained 'enormous' but accessible lives? Why not have a full-scale Darwin for the general market, the first with history-of-science credentials; a user-friendly Darwin, cheap, with a concealed apparatus, brief judicious quotations, a straightforward narrative, and above all a vivacious style?

I called for just such a biography in the early 1980s, years before Adrian's and my book was planned:

> What society today needs from the Darwin industry above all else
> . . . [is] a subtle, textured and thoroughly accessible portrait of
> Darwin that will command the respect of Darwin scholars, meet
> the demands of professional biographers and historians, and grip
> the imagination of the general reading public . . . The Darwin industry has only begun to show its ingenuity in contextuating Darwin's
> life, work and theories. Further progress will be indicated by the
> degree to which the pure waters of Darwin's science are muddied
> by the rich surrounding soil of political economy, natural theology,
> urban radicalism and provincial Dissent.[8]

[6] For an overview of the field, see Colp (1989).
[7] Bowler (1990), xi.
[8] Moore (1984), 19–20.

Adrian and I would write our book to such a brief. It would be a literary biography based on the latest research, a Darwin industry spin-off with broad consumer appeal.

Not that our main concern was to evade the critics of academic lives — far from it. There were powerful personal and economic forces that brought our *Darwin* into existence.

Preparing for 'Darwin'

I came to Britain in 1972 as a church history postgraduate expecting to write the life of a Victorian evangelical evolutionist. 'We don't like giving Americans Ph.D.s for biographies,' my supervisor scowled, so I changed tack and prepared a huge thesis on Protestant responses to Darwin. My interest in Darwin and religion persisted: I began to contemplate a biography of Darwin himself. A captain of the Darwin industry kindly warned me off: it would be a desiccated Darwin, he said — Darwin without the science. 'Spend ten years studying geology, zoology, palaeontology and botany, then reconsider'. Stymied again, I confined myself to researching Darwin's social and religious identity.

My quest for Darwin became highly personal. I fully grasped this only years later, after reading Richard Holmes's metabiographical masterpiece *Footsteps*. Holmes described a three-fold process of historical cognition that corresponded to my own.[9]

It began with a naive and only half-conscious *identification* with Darwin. I too was an ex-divinity student who had failed to enter the church and alienated his father by changing careers. I had visited South America, explored pampas and jungles, and wandered among the Andes. I had moved to London as a young man and fathered a daughter. My own father died in my absence, as Darwin's did, and so forth.

The next stage of my quest was marked by a strange intuition of Darwin's presence, a *haunting*. Committing what Holmes calls 'a deliberate act of psychological trespass', I began a series of 'personal essays' on Darwin's domesticity, affect for nature, religious views, and attitude to death. The research often made me feel as if he were with me, beside me, so close that I might even take part in his life. On occasion he seemed almost palpable, especially when I set out, pen and pad in hand, to follow his 'footsteps'.

One fresh spring morning in 1986 I reached Glen Roy in Scotland and

---

[9] Holmes (1987), 27, 66–9.

hiked up the hillside to the second great 'parallel road', or terrace, stretching round the glen. I found the very spot where Darwin himself sat in June 1838, trying to explain the enigmatic formation. I took his place, scribbling notes. It was a day like his, I imagined, with

> hot sun, broken cloud, gentle rush of stream below. Mottled green and lavender-brown hills – beautiful still, more if heather in bloom; snow melting fast . . .
>
> Wide green grazing valley; terraces very definite, spears of green piercing brown of old heather . . . sloping away from hill at perhaps 20° angle; . . . rolling contour as apparently contrived as a . . . roller coaster.
>
> Treeless except by stream. None where I stand except old gnarled one growing out of rocks, arguably the one in Darwin's engraving . . .
>
> Marvellous panorama: to east, on my right, the snow clad masses of Ben Nevis, Britain's highest peak, beyond which sea blue sky and clouds . . . Sweeping slopes down into Glen Roy – highway 150 years distant, it seems. Quiet – not a bird heard. Hills a cloud of solemn witnesses peering benignly, their strength & youth . . . spent. Gorge sweeps away curving to my left, terraces tracing along both sides continuously . . . grey single lane road, twisting crazily, like it . . . [did] . . . when CD was 29 years old.
>
> Was he moved by the poetic expanse before him, or merely by the professional determination to decode & explain? In this weather it is easy to understand how happy he was here. Alone, his own master, a summer stretching ahead, and life itself . . .
>
> A Land Rover hums into the picture, a visitor from another time and place; a . . . jet thunders distantly . . . Now the clouds are extinguishing the sun and I must leave – probably for ever. May these moments return to me if I should ever write Darwin's life.

This was a turning point. At Glen Roy I entered the third stage of my quest, Holmes's *complication*. The past became 'another country', inaccessible, when that jet cut through the sky and the Land Rover cruised up the glen. My identification with Darwin broke down and he vanished like a will-o'-the-wisp. I would never catch up with him or enter his world, let alone bring him back. I would have to recreate him 'using other sorts of skills and crafts and sensible magic'.

A further complication that discouraged any naive identification with

Darwin was my growing realisation that always and everywhere he gained his identity through social intercourse. At Glen Roy he might appear solitary, vulnerable, almost within romantic grasp; yet even here he had an eye on his geological mentors and his future wife. The gentleman-evolutionist was preparing to cloak himself in domesticity and retreat to a rural parish. Darwin was changing, England was changing – Victoria had just been crowned. His world excluded mine and threw me out of rapport. I had only his traces to go on, cryptic notes and texts. And – yes, still – a haunting sense of his presence, at the sites where he worked and lived.

Now the 'true biographic process' had begun, long before *Darwin* was started.

Besides personal motives there were brutal economic reasons for writing the book. In 1988 I moved unexpectedly to Cambridge, England from nearby Milton Keynes. Money was cheap and the housing market brisk after the Stock Exchange's 'Big Bang,' so I purchased property with a substantial bridging loan secured on my unsold flat. Like so many others in Thatcher's last months, I was to pay dearly for her fiscal policies. The economy began to overheat; interest rates soared and the housing market collapsed. My flat became an albatross, the loan a huge drain on my hard-pressed finances. By mid- 1989 I faced repossession or personal bankruptcy unless I found more cash for interest payments – fast. Fortunately, two London publishers had approached me to write about Darwin, with offers of an advance. I had put one on hold and was negotiating with the other when, on 7 June, I happened to phone Adrian to discuss the proofs of his *Politics of Evolution*.

We were distant old friends who kept in touch by commenting on each other's work. Adrian, an academic-turned-professional writer, told me that with *Politics* done he was pursuing new projects. He had actually started a Darwin biography but backed off when he heard that the field was full. His latest research was for a book on Irish republicanism. I protested at once, and not only about Adrian's loss to the history of science. Surely, I said, the world was big enough for several new Darwins. In fact, I myself had just been asked to write one, though I felt unqualified to tackle the details of Darwin's science. Then it hit me like an express train: Adrian had the right research background, in zoology and palaeontology; the last chapter of his *Politics*, 'Putting Darwin in the picture', was indebted to my 'personal essays'. 'So why not write a biography together?' I burst out spontaneously. The time was ripe, with Adrian free to start and my affairs in disarray. After a pregnant pause, Adrian replied quietly, 'You're on'.

We met weekly at Adrian's house in London to hammer out a proposal. Ours would be a people's Darwin – of this there was never any doubt. In July, before we could hawk the proposal around, Michael Joseph/Penguin got wind of it, and a contract suddenly arrived. We signed with a touch of trepidation, wondering whether a 'defiantly social' *Darwin* from the publishers of *Satanic Verses* would attract a *fatwa* first from philosophers or fundamentalists. Agents were assigned to sell foreign rights and by October Adrian was writing furiously. I joined him just before Christmas after taking leave from my university. We had twenty-four chapters mapped out and thirteen months to finish.

Making 'Darwin'

Our backgrounds and temperaments did not augur well for the collaboration. Those who knew us both said we were chalk and cheese: Adrian an Englishman, intensely private and self-contained, the master of collective entities and large-scale social forces; I a gregarious Yankee specialising on the individual's place in the world. Adrian had long avoided researching Darwin on the view that anyone so famous had to be hated first to be understood. I was a paid-up member of the Darwin industry. Adrian – in short – came at Darwin from the outside as an outsider, I approached Darwin personally and professionally from within.

But the differences between us turned out to be complementary, not contentious. Working feverishly, we had no time to ask why; in retrospect the reason seems obvious. At a deep, subarticulate level we trusted each other to be writing about the same Darwin. To Adrian he was primarily the Whig intellectual and gentleman-geologist – a respectable public figure. To me he was the failed ordinand and parish evolutionist – a rebellious private man. Together we grasped both sides of Darwin's paradoxical self, agreeing on why he feared being branded a 'Devil's Chaplain'. Nothing could have compensated for the lack of this basic unanimity. It was the *sine qua non* of our collaboration.

In writing, the division of labour followed the lines of our previous researches. Adrian dealt with the institutional and political aspects of Darwin's scientific work, including its technical details. I concentrated on the domestic and religious parts of Darwin's life, and on public reactions to his theories. The division of labour was never strict. Routinely we trespassed on each other, drafting whole pages. This was always welcome, for our historiographic aim was shared: to embed Darwin the man, his practices and theories, in a shifting social order.

Composition was chaotic. Each of us wrote first what he knew best, keen to make rapid progress. Adrian began at the beginning, with Darwin's Edinburgh years; I started with Darwin's death and burial in Westminster Abbey. Next we leapfrogged through the 1840s and 50s. I tackled Cambridge and the *Beagle* voyage at a trot while Adrian drafted Darwin's London period. I ploughed through the 1870s while Adrian laid down a patchwork of the 60s. Finally I wrote the first chapter; Adrian reworked it while I filled in his gaps, then together we revised the whole text one last time. It turned out to be twice the projected length. Split, subdivided, and recombined, the twenty-four chapters became forty-four.

Our working methods were more orderly, though as different as ourselves. Adrian overlaid screeds of notes directly from the sources to the pc screen, carving and polishing them into a coherent narrative. His hunt-and-pecking was a sight to behold, like Stradivari working wood. My own methods were primitive by contrast: skeleton outlines on screen, laboriously fleshed out and punched into shape, like a child's *papier-mâché*. Still, I churned out the chapters and exchanged them for Adrian's. Our disks flew like frisbees between Cambridge and London. We granted each other an editorial free hand. *Darwin*'s single style (often noted by reviewers) was the product of continuous, mutual revision.

The process involved tough negotiation. We prodded and pleaded with each other in hundreds of handwritten notes and in-text memos, some of which survive (addressed to the recipient).

> JIM My reworked Edinburgh chapter here. Vast chunks have been moved around. Still it's too cumbersome in places & requires your ministrations. It's hard to put the narrative in when — at this early period — there isn't any! What are we to do? Do what you like to lick this into shape.

> ADRIAN Your proposed rewording, 'Cambridge was a Christian society' is incredibly controversial and would get us into interminable arguments. To say that it was 'regarded' as such is much safer and indeed, I believe, incontrovertible.

> JIM Paley's turgid English has to be paraphrased into 1990s Penguin speak: be ruthless.

> ADRIAN You eliminated a most important element in my opening paragraphs: CD's attitude, now well formed, that people owed *him*

a visit, at his own convenience. I've been consciously laying the groundwork for his relations with Hooker, where the pattern is fully developed.

JIM Darwin's reference to the deist Herbert of Cherbury is so important and unexpected that it must have a direct reference after it! It's wonderful. Will 'Deism' be explained in the intro? I guess so. It would ruin the flow to do it here. This is a great revelation.

ADRIAN There's a dialogue between us going on in this chapter, more than in most for some reason. I've interpolated loads of explanations and self-justifications, while of course accepting most by far of your undoubted editorial improvements. Take my remarks for what they're worth and scrub them. All I want you to see is that, however poorly I express myself sometimes, there are usually interlocking reasons for certain forms of words. I struggle and fret to keep a breezy style without taking indecent liberties with the sources. On balance, sometimes, I go for accuracy rather than raciness. Also there is often a larger reason for introducing material, because it lays the groundwork for later chapters, or harks back to ones unwritten. Of course this reason may not be apparent as you read, nor should it be; otherwise we wouldn't be writing biography. But now I've explained myself, you might be brought to agree that some things may be more important than you thought.

JIM My prose is hackneyed — more subtlety needed.

ADRIAN Here's where you shine, 22 Sept.–19 Oct. 1832: Megatherium, Toxodon, Mylodon, and Glossotherium. Also (1) this is the place for a Lyellian update; (2) can you mention the dreadful dramatic days of 2–3 Oct. when he was marooned ashore? — this was on the first anniversary of his leaving The Mount! (3) note FitzRoy's curt remark apropos all the bones.

JIM This chapter is revamped. About ½ page of superfluous material on barnacles removed for streamlining. Also your Norton/Sterling material slimmed down & moved. Reason for moving is that Dr Darwin died on p. 15. He is stiff by the end of the page; tears everywhere. Then on 16 you have him dying! And by 17 he is alive again. Dead is dead, so I took all CD's religious reading prior to the Doctor's death & reworked it into the 1848 period — and kept CD's eye on his father, an unbeliever, slipping away as he read. Okay?

It was. We grew to trust each other's editorial judgement even as we grappled together like exhausted swimmers, desperate not to drown. On 22 November 1990, the week my wretched flat finally sold (leaving me still five-figures in debt) and just a fortnight after the birth of his first child, Adrian sent comfort: 'JIM How are you holding up? These are difficult times, always, just at the end of a mammoth & herculean task like writing a book. Beats me why anybody ever writes a second (as it does that anyone could go thru labour twice)! Keep going – we're almost there'.

We weren't. There were six months to go, 14 chapters to write. The publisher's noose tightened by the week. Adrian's baby kept him up all hours when the book didn't. On my own, I was having technicolour nightmares. In April 1991 came the ultimatum: a despatch rider would be sent round for the manuscript on such-and-such a day. No manuscript, no book. Publication would be put off a year. By now Adrian's partner was back at work and he was a full-time house parent, with only evenings and weekends to write. I took up the slack, frantic at the economic consequences of delay.

Somehow we finished. The disks went off on 19 May and bound proofs arrived within a fortnight. We almost fell out compiling the index, our bottled-up tensions on the brink. But again we met the deadline, and in August our genial, long-suffering editors, Susan Watt and Anne Askuith, sent handsome copies of *Darwin*, followed by congratulatory champagne. We cooled down.

Some working assumptions

With deadlines looming we had little chance to ponder our presuppositions. Neither of us had a 'theory' of biography, nor did we model our *Darwin* on another. I myself cracked a modern life of Darwin only twice in eighteen months – Peter Brent's. Like aviators without an altimeter, we flew by 'the seat of our pants', dodging obstacles in the fog, hoping to touch down safely before we ran out of time.

The individualistic bias of biography was the first problem we had to surmount. Our book was to be about a man, but – as I had long since realised – his identity was socially acquired. His father's family, the Glutton Club, the *Beagle* men, the Whig intellectuals, the Kentish squirearchy, the professional scientists, his own family – one after another these groups passed into the foreground or the background of Darwin's life. He was never the 'solitary brute' he made himself out to be, nor was he static. He moved constantly, sealing and or severing relationships, changing his society in a society that itself was changing. To follow him Adrian and I had to keep 'pulling focus'. Our narrative brought this group or that clearly into frame, sometimes with Darwin visible, sometimes not. We tracked his movements through the larger world only to land him in the petty complexities of Downe parish. There too he was never alone, even in death. His body was snatched from grieving loved ones for the greater glory of the nation's new professionals.

We also puzzled over the tension between teleology and contingency in our narrative. Real lives have a beginning and an end, between which a story line imparts a vector. Yet in a good story there are surprises – things literally happen. Take the *Beagle* voyage. We now know that Darwin survived it, sorted out the Galapagos finches, and made straight for evolution. With hindsight his five years at sea was the turning point of his career. The 1970s BBC television series, 'The Voyage of Charles Darwin,' made this view explicit. Each programme begins with Darwin reading from his autobiography, written forty years later. The scope for surprise was limited.

But suppose Uncle Jos Wedgwood had not convinced the Doctor that Charles should go. Or FitzRoy the physiognomist had turned up his nose at Darwin's. Or Fanny Owen, her engagement broken, had bumped into her heartbroken old flame in Plymouth, just before he sailed. Or a fourth roller had swamped the *Beagle* at Cape Horn. Or FitzRoy had not come to his senses at Valparaiso and Darwin had jumped ship. What if he had yielded to his impulse to return at any point?

The contingencies were real, as Darwin himself believed. His prospects were always more or less uncertain, depending on where he was in the world. *Darwin* recaptures that uncertainty without losing narrative thrust. We cloaked our story with contingencies, even while laying the groundwork for future developments. On the voyage Darwin had to collect fossils and finches, corals and conch shells, for use in later chapters. He had to be shaken by 'savage' Fuegians in order to countenance human evolution back in London. Such hindsight is a biographer's prerogative but ours does not advertise itself – it is tacit.

Chronology was a problem for us early on. We soon decided for the readers' sake to keep the story going in a straight line – few flashbacks, rare recaps, no historiographic asides. When Darwin fades momentarily from the picture – for instance, only one 1827 letter survives, written *to* him on 31 December – we pull focus, describing events from a friend or relative's point of view. Or we sketch in the social context, plunge into politics, use props or improvise. Inferences help to fill the gaps — thus an outbound journey must have a return — as do unfalsifiable commonplaces. People eat, sleep, make love, get sick, and die. Darwin himself was mainly sick. We found the narrative possibilities endless.

And what of the narrative itself? Socialised not individualistic, contingent and teleological at once, proceeding diachronically, it still might have featured the first person, like so many academic lives. But no — Adrian and I chose to vanish into the credits. We adopted a 'ciné theory' of narration. Just as in a cinema the technical work of film-making is hidden, and the projection goes on silently, unobtrusively, outside the draped and darkened room, so *Darwin* offers a 'space' where readers may suspend disbelief and enter another world, undistracted by the machinery of scholarship. The textual sights and sounds are contrived no less than a feature film's; the authorial voice, like a screenwriter's, is always heard, even in the actors. But the voice does not speak for itself. Biographers are like ventriloquists, reanimating the dead, making dry texts live.

Our narrative is dense, 'thick'. We mixed everything together, from aborigines to zoos, reinvesting Darwin's science with real-life complexity. This involved a 'cake-bake theory' of composition. The making of *Darwin* was an irreversible, hi-entropic process: from simple, discrete ingredients we concocted a full, rich and detailed story. Each chapter became a generous slice of life, a tasty read. Extracting the pure 'science' now would be like trying to unbake a cake — impossible.

Speaking of sweets reminds me again of Charlie's chocolate factory and its cast — Augustus Gloop, Veruca Salt, Violet Beauregarde, Mike Teavee, and not least Mr Wonka and the Oompa-Loompas. We recall these names today because Roald Dahl threaded them into a vivid, engaging story. If in future readers of our *Darwin* remember Robert Grant, Fanny Owen, Joseph Hooker, Ernst Haeckel, and not least Jemmy Button and the Fuegians, it may be because Adrian and I brought them to life in another Charlie's epic.

But *Darwin* should not remain memorable for this alone. (And what's 'memorable' is all that counts in history, according to *1066 and All That*). A *Times Higher* reviewer asked whether our 'conventional biography' is 'the best vehicle for promoting important historiographical points'. *Darwin*, he suggested, is both opportunist and subversive. 'In their desire to penetrate the market for conventional biographies, the authors have been forced to adopt a format that will leave many nonspecialist readers unaware of the fundamental issues buried in the book's wealth of social detail'. The problem is that *Darwin* doesn't nail its 'colours . . . to the mast' through an 'explicit analysis' of academic debates.[10]

Of course. General readers do not have to know about the 'fundamental issues' in Darwin historiography (or they can read Peter Bowler's recent book if they need to). For a century biographers have omitted to enlighten them, and not a critic has wagged his tongue. The public's image of Darwin as a saintly scientist and intellectual revolutionary has always been taken in by osmosis, from television, textbooks and the popular press. I see no reason why this process should not continue. Our *Darwin* offers 'a defiantly social portrait': the countervailing image of a respectable man with a dangerous theory, picking his way through a menacing world. It is a memorable image, and it need not come in plain brown paper wrappers or bear a skull-and-crossbones. Let the people read, let the image sink in, let other Darwins bloom. Then we shall see what the 'best vehicle' is for 'promoting important historiographical points'.

Biography historicises. It may yet prove to be the most effective way of informing the widest audience about the politics of scientific practice and the cultural formation of natural knowledge.

Bibliography

Blow, S. (1987) Thumping great lives. *The Times*, 17 October, 10.

Bowler, P.J. (1990) *Charles Darwin: The Man and His Influence*. Oxford: Blackwell.

Bowler, P.J. (1991) Struggling to survive and adapt. *Times Higher Education Supplement*, 4 Oct., 24.

Colp, R., Jr (1989) Charles Darwin's past and future biographers. *History of Science*, 27, 167–97.

Holmes, R. (1987) *Footsteps: Adventures of a Romantic Biographer*. Harmondsworth: Penguin.

Judd, A. (1991) Bert into D. H. *Times Literary Supplement*, 13 Sept., 12–13.

Moore, J. (1984) On revolutionizing the Darwin industry: a centennial retrospect. *Radical Philosophy*, 37, 19–20.

Neve, M. (1991) The long argument: Darwin and the economy of nature. *Times Literary Supplement*, 13 September, 3–4.

Williams, R. (1973) Base and superstructure in Marxist cultural theory. *New Left Review*, 82, 3–16.

[10] Bowler (1991).

Index

Page numbers in *italics* refer to illustrations